HUTCHINSON POCKET

Guide to
# Countries
of the World

**Other titles in the Hutchinson Pocket series:**

Pocket Dictionary of Biology
Pocket Dictionary of Chemistry
Pocket Dictionary of Geography
Pocket Dictionary of Mathematics
Pocket Dictionary of Physics
Pocket Dictionary of Science
Pocket Dictionary of 20th Century World History

*Published October 1993:*

Pocket Dictionary of Computing
Pocket Dictionary of Quotations
Pocket Quiz Book
Pocket Chronology of World History

# HUTCHINSON POCKET

# Guide to Countries of the World

Helicon

Copyright © Helicon Publishing Ltd 1993

All rights reserved

Helicon Publishing Ltd
42 Hythe Bridge Street
Oxford OX1 2EP

Printed and bound in Great Britain by
Unwin Brothers Ltd, Old Woking, Surrey

ISBN 0 09 178270 8

**British Cataloguing in Publication Data**

A catalogue record for this book is available
from the British Library

**Editorial director**
Michael Upshall

**Project editor**
Sara Jenkins-Jones

**Text editor**
Catherine Thompson

**Art editor**
Terence Caven

**Page make-up**
Helen Bird

**Production**
Tony Ballsdon

# Afghanistan
Republic of
(*Jamhuria Afghanistan*)

*area* 652,090 sq km/251,707 sq mi
*capital* Kabul
*towns* Kandahar, Herat, Mazar-i-Sharif
*physical* mountainous in centre and NE, plains in N and SW
*environment* an estimated 95% of the urban population is without access to sanitation services
*head of state* Burhanuddin Rabbani from 1992
*head of government* to be announced
*political system* emergent democracy
*exports* dried fruit, natural gas, fresh fruits, carpets; small amounts of rare minerals, karakul lamb skins, and Afghan coats
*currency* afgháni
*population* (1989) 15,590,000 (more than 5 million became refugees after 1979); growth rate 0.6% p.a.
*life expectancy* (1986) men 43, women 41
*languages* Pushtu, Dari (Persian)
*religion* Muslim (80% Sunni, 20% Shi'ite)
*literacy* men 39%, women 8% (1985 est)
*GNP* $3.3 bn (1985); $275 per head

*chronology*
**1747** Afghanistan became an independent emirate.
**1839–42 and 1878–80** Afghan Wars instigated by Britain to counter the threat to British India from expanding Russian influence in Afghanistan.
**1919** Afghanistan recovered full independence following Third Afghan War.
**1953** Lt-Gen Daud Khan became prime minister and introduced reform programme.
**1963** Daud Khan forced to resign and constitutional monarchy established.
**1973** Monarchy overthrown in coup by Daud Khan.
**1978** Daud Khan ousted by Taraki and the Democratic Party of Afghanistan (PDPA).
**1979** Taraki replaced by Hafizullah Amin; Soviet Union entered country to prop up government; they installed Babrak Karmal in power. Amin executed.
**1986** Replacement of Karmal as leader by Dr Najibullah Ahmadzai. Partial Soviet troop withdrawal.
**1988** New non-Marxist constitution adopted.
**1989** Complete withdrawal of Soviet troops; state of emergency imposed in response to intensification of civil war.
**1991** UN peace plan accepted by President Najibullah but rejected by the mujaheddin. US and Soviet military aid withdrawn. Mujaheddin began talks with Russians and Kabul government.
**1992** April: Najibullah regime overthrown. June: after a succession of short-term presidents, Burhanuddin Rabbani named interim head of state; Islamic law introduced. Sept: Hezb-i-Islami barred from government participation after shell attacks on Kabul. Dec: Rabbani elected president for two-year term by constituent assembly.
**1993** Jan: renewed bombardment of Kabul by Hezb-i-Islami and other rebel forces. Interim parliament appointed by constituent assembly. March: peace agreement signed between Rabbani and dissident mujaheddin leader Gulbuddin Hekmatyar, under which Hekmatyar to become prime minister.

## Albania
Republic of
(*Republika e Shqipërisë*)

*area* 28,748 sq km/11,097 sq mi
*capital* Tiranë
*towns* Shkodër, Elbasan, Vlorë, Durrës
*physical* mainly mountainous, with rivers flowing E–W, and a narrow coastal plain
*head of state* Sali Berisha from 1992
*head of government* Alexander Meksi from 1992
*political system* emergent democracy
*exports* crude oil, bitumen, chrome, iron ore, nickel, coal, copper wire, tobacco, fruit, vegetables
*currency* lek
*population* (1990 est) 3,270,000; growth rate 1.9% p.a.
*life expectancy* men 69, women 73
*languages* Albanian, Greek
*religion* Muslim 70%, although all religion banned 1967–90
*literacy* 75% (1986)
*GNP* $2.8 bn (1986 est); $900 per head

*chronology*
*c. 1468* Albania made part of the Ottoman Empire.
*1912* Independence achieved from Turkey.
*1925* Republic proclaimed.
*1928–39* Monarchy of King Zog.
*1939–44* Under Italian and then German rule.
*1946* Communist republic proclaimed under the leadership of Enver Hoxha.
*1949* Admitted into Comecon.
*1961* Break with Khrushchev's USSR.
*1967* Albania declared itself the 'first atheist state in the world'.
*1978* Break with 'revisionist' China.
*1985* Death of Hoxha.
*1987* Normal diplomatic relations restored with Canada, Greece, and West Germany.
*1988* Attendance of conference of Balkan states for the first time since the 1930s.
*1990* One-party system abandoned; first opposition party formed.
*1991* April: Party of Labour of Albania (PLA) won first multiparty elections; Ramiz Alia re-elected president; three successive governments formed. PLA renamed Socialist Party of Albania (PSS).
*1992* Former communist officials arrested on corruption charges. Presidential elections won by Democratic Party of Albania (DP); Sali Berisha elected president. Alia charged with corruption and abuse of power; totalitarian and communist parties banned.
*1993* Jan: Nexhmije Hoxha, widow of Enver Hoxha, sentenced to nine years' imprisonment for misuse of government funds 1985–90.

# Algeria

Democratic and Popular Republic of (*al-Jumhuriya al-Jazairiya ad-Dimuqratiya ash-Shabiya*)

*area* 2,381,741 sq km/919,352 sq mi
*capital* al-Jazair (Algiers)
*towns* Qacentina/Constantine, Ouahran/Oran, Annaba/Bône
*physical* coastal plains backed by mountains in N; Sahara desert in S
*head of state* Ali Kafi from 1992
*head of government* Belnid Absessalem from 1992
*political system* semi-military rule
*exports* oil, natural gas, iron, wine, olive oil
*currency* dinar
*population* (1990 est) 25,715,000 (83% Arab, 17% Berber); growth rate 3.0% p.a.
*life expectancy* men 59, women 62
*languages* Arabic (official); Berber, French
*religion* Sunni Muslim (state religion)
*literacy* men 63%, women 37% (1985 est)
*GDP* $64.6 bn; $2,796 per head
*chronology*
*1962* Independence achieved from France. Republic declared. Ahmed Ben Bella elected prime minister.

# Algeria

*1963* Ben Bella elected Algeria's first president.
*1965* Ben Bella deposed by military, led by Colonel Houari Boumédienne.
*1976* New constitution approved.
*1978* Death of Boumédienne.
*1979* Benjedid Chadli elected president. Ben Bella released from house arrest. National Liberation Front (FLN) adopted new party structure.
*1981* Algeria helped secure release of US prisoners in Iran.
*1983* Chadli re-elected.
*1988* Riots in protest at government policies; 170 killed. Reform programme introduced. Diplomatic relations with Egypt restored.
*1989* Constitutional changes proposed, leading to limited political pluralism.
*1990* Fundamentalist Islamic Salvation Front (FIS) won Algerian municipal and provincial elections.
*1991* Dec: FIS won first round of multiparty elections.
*1992* Jan: Chadli resigned; military took control of government; Mohamed Boudiaf became president; FIS leaders detained. Feb: state of emergency declared. March: FIS ordered to disband. June: Boudiaf assassinated; Ali Kafi chosen as new head of state and Belnid Absessalem as prime minister.

# Andorra
Principality of
(*Principat d'Andorra*)

*area* 468 sq km/181 sq mi
*capital* Andorra-la-Vella
*towns* Les Escaldes
*physical* mountainous, with narrow valleys
*heads of state* Joan Marti i Alanis (bishop of Urgel, Spain) and François Mitterrand (president of France)
*head of government* Oscar Riba Reig from 1989
*political system* semi-feudal co-principality
*exports* main industries are tourism and tobacco
*currency* French franc and Spanish peseta
*population* (1990) 51,000 (30% Andorrans, 61% Spanish, 6% French)
*languages* Catalan (official); Spanish, French
*religion* Roman Catholic
*literacy* 100% (1987)
*GDP* $300 million (1985)
*chronology*
*1278* Treaty signed making Spanish bishop and French count joint rulers of Andorra (the king of France later inherited the count's right).

*1970* Extension of franchise to third-generation female and second-generation male Andorrans.
*1976* First political organization (Democratic Party of Andorra) formed.
*1977* Franchise extended to first-generation Andorrans.
*1981* First prime minister appointed by General Council.
*1982* With the appointment of an Executive Council, executive and legislative powers were separated.
*1991* Andorra's first constitution planned; links with European Community (EC) formalized.
*1993* First constitution approved in a referendum.

# Angola
People's Republic of
(*República Popular de Angola*)

*area* 1,246,700 sq km/481,226 sq mi
*capital* and chief port Luanda
*towns* Lobito and Benguela, Huambo, Lubango
*physical* narrow coastal plain rises to vast interior plateau with rainforest in NW; desert in S
*head of state and government* José Eduardo dos Santos from 1979
*political system* socialist republic
*exports* oil, coffee, diamonds, palm oil, sisal, iron ore, fish
*currency* kwanza
*population* (1989 est) 9,733,000 (largest ethnic group Ovimbundu); growth rate 2.5% p.a.
*life expectancy* men 40, women 44
*languages* Portuguese (official); Bantu dialects
*religions* Roman Catholic 68%, Protestant 20%, animist 12%
*literacy* 20%
*GDP* $2.7 bn; $432 per head

# Angola

*recent chronology*

**1951** Angola became an overseas territory of Portugal.
**1956** First independence movement formed, the People's Movement for the Liberation of Angola (MPLA).
**1961** Unsuccessful independence rebellion.
**1962** National Front for the Liberation of Angola (FNLA) formed.
**1966** National Union for the Total Independence of Angola (UNITA) formed.
**1975** Independence achieved from Portugal. Transitional government formed from representatives of MPLA, FNLA, UNITA, and Portuguese government. MPLA proclaimed People's Republic of Angola under the presidency of Dr Agostinho Neto. FNLA and UNITA proclaimed People's Democratic Republic of Angola.
**1976** MPLA gained control of most of the country. South African troops withdrawn, but Cuban units remained.
**1977** MPLA restructured to become the People's Movement for the Liberation of Angola–Workers' Party (MPLA–PT).
**1979** Death of Neto; succeeded by José Eduardo dos Santos.
**1980** Constitution amended to provide for an elected people's assembly. UNITA guerrillas, aided by South Africa, continued raids against the Luanda government and bases of the South West Africa People's Organization (SWAPO) in Angola.
**1984** South Africa promised to withdraw its forces if the Luanda government guaranteed that areas vacated would not be filled by Cuban or SWAPO units (the Lusaka Agreement).
**1985** South African forces officially withdrawn.
**1986** Further South African raids into Angola. UNITA continuing to receive South African support.
**1988** Peace treaty, providing for the withdrawal of all foreign troops, signed with South Africa and Cuba.
**1989** Cease-fire with UNITA broke down; guerrilla activity resumed
**1990** Peace offer by rebels. Return to multiparty politics promised.
**1992** MPLA–PT's general-election victory fiercely disputed by UNITA, plunging the country into renewed civil war. UNITA accepted seats in the new government, but fighting continued.
**1993** Continued fighting posed the threat of return to civil war.

# Antigua and Barbuda
State of

*area* Antigua 280 sq km/108 sq mi, Barbuda 161 sq km/62 sq mi, plus Redonda 1 sq km/0.4 sq mi
*capital* and chief port St John's
*towns* Codrington (on Barbuda)
*physical* low-lying tropical islands of limestone and coral with some higher volcanic outcrops; no rivers and low rainfall result in frequent droughts and deforestation
*head of state* Elizabeth II from 1981 represented by governor general
*head of government* Vere C Bird from 1981
*political system* liberal democracy
*exports* sea-island cotton, rum, lobsters
*currency* Eastern Caribbean dollar
*population* (1989) 83,500; growth rate 1.3% p.a.
*life expectancy* 70 years
*language* English
*religion* Christian (mostly Anglican)
*literacy* 90% (1985)
*GDP* $173 million (1985); $2,200 per head
*chronology*

# Antigua and Barbuda

**1493** Antigua visited by Christopher Columbus.
**1632** Antigua colonized by English settlers.
**1667** Treaty of Breda formally ceded Antigua to Britain.
**1871–1956** Antigua and Barbuda administered as part of the Leeward Islands federation.
**1967** Antigua and Barbuda became an associated state within the Commonwealth, with full internal independence.
**1971** Progressive Labour Movement (PLM) won the general election by defeating the Antigua Labour Movement (ALP).
**1976** PLM called for early independence, but ALP urged caution. ALP won the general election.
**1981** Independence from Britain achieved.
**1983** Assisted US invasion of Grenada.
**1984** ALP won a decisive victory in the general election.
**1985** ALP re-elected.
**1989** Another sweeping general election victory for the ALP under Vere Bird.
**1991** Bird remained in power despite calls for his resignation.

## Argentina
Republic of
(*República Argentina*)

*area* 2,780,092 sq km/1,073,116 sq mi
*capital* Buenos Aires (to move to Viedma)
*towns* Rosario, Córdoba, Tucumán, Mendoza, Santa Fé, La Plata
*physical* mountains in W, forest and savanna in N, pampas (treeless plains) in E central area, Patagonian plateau in S; rivers Colorado, Salado, Paraná, Uruguay, Río de la Plata estuary
*territories* part of Tierra del Fuego
*environment* an estimated 20,000 sq km/7,700 sq mi of land has been swamped with salt water
*head of state and government* Carlos Menem from 1989
*political system* emergent democratic federal republic
*exports* livestock products, cereals, wool, tannin, peanuts, minerals (coal, copper, molybdenum, gold, silver, lead, zinc, barium, uranium); huge resources of oil, natural gas, hydroelectric power
*currency* peso = 10,000 australs (which it replaced 1992)
*population* (1990 est) 32,686,000; growth rate 1.5% p.a.

# Argentina

*life expectancy* men 66, women 73
*languages* Spanish (official); English, Italian, German, French
*religion* Roman Catholic (state-supported)
*literacy* men 96%, women 95% (1985 est)
*GDP* $70.1 bn (1990); $2,162 per head

*chronology*
*1816* Independence achieved from Spain, followed by civil wars.
*1946* Juan Perón elected president, supported by his wife 'Evita'.
*1952* 'Evita' Perón died.
*1955* Perón overthrown and civilian administration restored.
*1966* Coup brought back military rule.
*1973* Perón returned from exile in Spain as president, with his third wife, Isabel, as vice president.
*1974* Perón died, succeeded by Isabel.
*1976* Coup resulted in rule by a military junta led by Lt-Gen Jorge Videla. Congress dissolved, and hundreds of people detained.
*1976–83* Ferocious campaign ('dirty war') against left-wing elements.
*1978* Videla retired. Succeeded by General Roberto Viola, who promised a return to democracy.
*1981* Viola died suddenly. Replaced by General Leopoldo Galtieri.
*1982* With a deteriorating economy, Galtieri sought popular support by ordering an invasion of the British-held Falkland Islands. After losing the short war, Galtieri was replaced by General Reynaldo Bignone.
*1983* Amnesty law passed and democratic constitution of 1853 revived. General elections won by Raúl Alfonsín and his party. Armed forces under scrutiny.
*1984* National Commission on the Disappearance of Persons (CONADEP) reported on over 8,000 people who had disappeared during the 'dirty war' of 1976–83.
*1985* A deteriorating economy forced Alfonsín to seek help from the International Monetary Fund and introduce an austerity programme.
*1988* Unsuccessful army coup.
*1989* Carlos Menem, of the Justicialist Party, elected president.
*1990* Full diplomatic relations with the UK restored. Menem elected Justicialist Party leader. Revolt by army officers thwarted.
*1992* New currency introduced.

## Armenia
Republic of

*area* 29,800 sq km/11,500 sq mi
*capital* Yerevan
*towns* Kumayri (formerly Leninakan)
*physical* mainly mountainous (including Mount Ararat), wooded
*head of state* Levon Ter-Petrossian from 1990
*head of government* Gagik Arutyunyan from 1991
*political system* emergent democracy
*products* copper, molybdenum, cereals, cotton, silk
*population* (1991) 3,580,000 (90% Armenian, 5% Azeri, 2% Russian, 2% Kurd)
*language* Armenian
*religion* traditionally Armenian Christian
*chronology*
*1918* Became an independent republic.
*1920* Occupied by the Red Army.
*1936* Became a constituent republic of the USSR.
*1988* Feb: demonstrations in Yerevan called for transfer of Nagorno-Karabakh from Azerbaijan to Armenian control. Dec: earthquake claimed around 25,000 lives and caused extensive damage.
*1989* Jan–Nov: strife-torn Nagorno-Karabakh placed under

'temporary' direct rule from Moscow. Pro-autonomy Armenian National Movement founded. Nov: civil war erupted with Azerbaijan over Nagorno-Karabakh.
*1990* March: Armenia boycotted USSR constitutional referendum. Aug: nationalists secured control of Armenian supreme soviet; former dissident Levon Ter-Petrossian indirectly elected president; independence declared. Nakhichevan republic affected by Nagorno-Karabakh dispute.
*1991* March: overwhelming support for independence in referendum. Dec: Armenia joined new Commonwealth of Independent States (CIS); Nagorno-Karabakh declared its independence; Armenia granted diplomatic recognition by USA.
*1992* Jan: admitted into Conference on Security and Cooperation in Europe (CSCE). March: joined United Nations (UN). Conflict over Nagorno-Karabakh worsened.

**Australia**
Commonwealth of

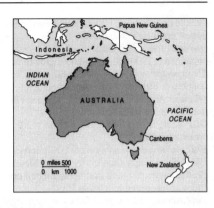

*area* 7,682,300 sq km/2,966,136 sq mi
*capital* Canberra
*towns* Adelaide, Brisbane, Darwin, Melbourne, Perth, Sydney
*physical* world's smallest, flattest, and driest continent (40% lies in the tropics, one-third is desert, and one-third is marginal grazing); Great Sandy, Gibson, Great Victoria, and Simpson deserts; the Great Barrier Reef (largest coral reef in the world); Great Dividing Range and Australian Alps in the E. The fertile SE region is watered by the Darling, Lachlan, Murrumbridgee, and Murray rivers; rivers in the interior are seasonal. Lake Eyre basin and Nullarbor Plain in the S
*territories* Norfolk Island, Christmas Island, Cocos (Keeling) Islands, Ashmore and Cartier Islands, Coral Sea Islands, Heard Island and McDonald Islands, Australian Antarctic Territory
*environment* an estimated 75% of Australia's northern tropical rainforest has been cleared for agriculture or urban development since Europeans first settled there in the early 19th century
*head of state* Elizabeth II from 1952, represented by governor general
*head of government* Paul Keating from 1991
*political system* federal constitutional monarchy
*exports* world's largest exporter of sheep, wool, diamonds, alumina, coal, lead and refined zinc ores, and mineral sands; other exports

include cereals, beef, veal, mutton, lamb, sugar, nickel (world's second largest producer), iron ore
*currency* Australian dollar
*population* (1990 est) 16,650,000; growth rate 1.5% p.a.
*life expectancy* men 75, women 80
*languages* English, Aboriginal languages
*religions* Anglican 26%, other Protestant 17%, Roman Catholic 26%
*literacy* 98.5.% (1988)
*GDP* $220.96 bn (1988); $14,458 per head
*recent chronology*
*1942* Achieved autonomy from UK in internal and external affairs.
*1944* Liberal Party founded by Robert Menzies.
*1966* Menzies resigned after being Liberal prime minister for 17 years, and was succeeded by Harold Holt.
*1967* Aborigines achieved full citizenship rights.
*1968* John Gorton became prime minister after Holt's death.
*1971* Gorton succeeded by William McMahon, heading a Liberal–Country Party coalition.
*1972* Gough Whitlam became prime minister, leading a Labor government.
*1975* Senate blocked the government's financial legislation; Whitlam dismissed by the governor general, who invited Malcolm Fraser to form a Liberal–Country Party caretaker government. The action of the governor general, John Kerr, was widely criticized.
*1978* Northern Territory attained self-government.
*1983* Australian Labor Party returned to power under Bob Hawke.
*1986* Australia Act passed by UK government, eliminating last vestiges of British legal authority in Australia.
*1988* Labor foreign minister Bill Hayden appointed governor general designate. Free-trade agreement with New Zealand signed.
*1990* Hawke won record fourth election victory, defeating Liberal Party by small majority.
*1991* Paul Keating became new Labor Party leader and prime minister.
*1992* Keating's popularity declined as economic problems continued. Oath of allegiance to British Crown abandoned.
*1993* Labor Party won general election, entering fifth term of office.

# Austria

Republic of
(*Republik Österreich*)

*area* 83,500 sq km/32,374 sq mi
*capital* Vienna
*towns* Graz, Linz, Salzburg, Innsbruck
*physical* landlocked mountainous state, with Alps in W and S and low relief in E where most of the population is concentrated
*environment* Hainburg, the largest primeval forest left in Europe, under threat from a dam project (suspended 1990)
*head of state* Thomas Klestil from 1992
*head of government* Franz Vranitzky from 1986
*political system* democratic federal republic
*exports* lumber, textiles, clothing, iron and steel, paper, machinery and transport equipment, foodstuffs
*currency* schilling
*population* (1990 est) 7,595,000; growth rate 0.1% p.a.
*life expectancy* men 70, women 77
*language* German
*religions* Roman Catholic 85%, Protestant 6%
*literacy* 98% (1983)

***GDP*** $183.3 bn (1987); $11,337 per head

*chronology*

**1867** Emperor Franz Josef established dual monarchy of Austria–Hungary.

**1914** Archduke Franz Ferdinand assassinated by a Serbian nationalist; Austria–Hungary invaded Serbia, precipitating World War I.

**1918** Habsburg empire ended; republic proclaimed.

**1938** Austria incorporated into German Third Reich by Hitler (the *Anschluss*).

**1945** Under Allied occupation, constitution of 1920 reinstated and coalition government formed by the Socialist Party of Austria (SPÖ) and the Austrian People's Party (ÖVP).

**1955** Allied occupation ended, and the independence of Austria formally recognized.

**1966** ÖVP in power with Josef Klaus as chancellor.

**1970** SPÖ formed a minority government, with Dr Bruno Kreisky as chancellor.

**1983** Kreisky resigned and was replaced by Dr Fred Sinowatz, leading a coalition.

**1986** Dr Kurt Waldheim elected president. Sinowatz resigned, succeeded by Franz Vranitzky. No party won an overall majority; Vranitzky formed a coalition of the SPÖ and the ÖVP, with ÖVP leader, Dr Alois Mock, as vice chancellor.

**1989** Austria sought European Community membership.

**1990** Vranitzky re-elected.

**1991** Bid for EC membership endorsed by the Community.

**1992** Thomas Klestil elected president, replacing Waldheim.

# Azerbaijan
Republic of

*area* 86,600 sq km/33,400 sq mi
*capital* Baku
*towns* Gyandzha (formerly Kirovabad), Sumgait
*physical* Caspian Sea; the country ranges from semidesert to the Caucasus Mountains
*head of state* Albulfaz Elchibey from 1992
*head of government* to be appointed
*political system* emergent democracy
*products* oil, iron, copper, fruit, vines, cotton, silk, carpets
*population* (1990) 7,145,600 (83% Azeri, 6% Russian, 6% Armenian)
*language* Turkic
*religion* traditionally Shi'ite Muslim
*chronology*
*1917–18* A member of the anti-Bolshevik Transcaucasian Federation.
*1918* Became an independent republic.
*1920* Occupied by the Red Army.
*1922–36* Formed part of the Transcaucasian Federal Republic with Georgia and Armenia.
*1936* Became a constituent republic of the USSR.
*1988* Riots followed Nagorno-Karabakh's request for transfer to Armenia.

***1989*** Jan–Nov: strife-torn Nagorno-Karabakh placed under 'temporary' direct rule from Moscow. Azerbaijan Popular Front established. Nov: civil war erupted with Armenia.
***1990*** Jan: Soviet troops dispatched to Baku to restore order. Aug: communists won parliamentary elections. Nakhichevan republic affected by Nagorno-Karabakh dispute.
***1991*** Aug: Azeri leadership supported attempted anti-Gorbachev coup in Moscow; independence declared. Sept: former communist Ayaz Mutalibov elected president. Dec: joined new Commonwealth of Independent States (CIS); Nagorno-Karabakh declared independence.
***1992*** Jan: admitted into Conference on Security and Cooperation in Europe (CSCE); March: Mutalibov resigned; Azerbaijan became a member of the United Nations (UN); accorded diplomatic recognition by the USA. June: Albulfaz Elchibey, leader of the Popular Front, elected president; renewed campaign against Armenia in the fight for Nagorno-Karabakh.
***1993*** Prime Minister Rakham Guseinov resigned over differences with President Elchibey.

# Bahamas
Commonwealth of the

*area* 13,864 sq km/5,352 sq mi
*capital* Nassau on New Providence
*towns* Alice Town, Andros Town, Hope Town, Spanish Wells, Freeport, Moss Town
*physical* comprises 700 tropical coral islands and about 1,000 cays
*principal islands* Andros, Grand Bahama, Great Abaco, Eleuthera, New Providence, Berry Islands, Biminis, Great Inagua, Acklins, Exumas, Mayaguana, Crooked Island, Long Island, Cat Island, Rum Cay, Watling (San Salvador) Island
*head of state* Elizabeth II from 1973, represented by governor general
*head of government* Hubert Ingraham from 1992
*political system* constitutional monarchy
*exports* cement, pharmaceuticals, petroleum products, crawfish, salt, aragonite, rum, pulpwood; over half the islands' employment comes from tourism
*currency* Bahamian dollar
*population* (1990 est) 251,000; growth rate 1.8% p.a.
*life expectancy* men 67, women 74
*languages* English and some Creole

*religions* 29% Baptist, 23% Anglican, 22% Roman Catholic
*literacy* 95% (1986)
*GDP* $2.7 bn (1987); $11,261 per head
*chronology*
*1964* Independence achieved from Britain.
*1967* First national assembly elections; Lynden Pindling became prime minister.
*1972* Constitutional conference to discuss full independence.
*1973* Full independence achieved.
*1983* Allegations of drug trafficking by government ministers.
*1984* Deputy prime minister and two cabinet ministers resigned. Pindling denied any personal involvement and was endorsed as party leader.
*1987* Pindling re-elected despite claims of frauds.
*1992* Free National Movement (FNM) led by Hubert Ingraham won absolute majority in assembly elections.

## Bahrain
State of
(*Dawlat al Bahrayn*)

*area* 688 sq km/266 sq mi
*capital* Manama on the largest island (also called Bahrain)
*towns* Muharraq, Jidd Hafs, Isa Town, Mina Sulman
*physical* 35 islands, composed largely of sand-covered limestone; generally poor and infertile soil; flat and hot
*environment* a wildlife park on Bahrain preserves the endangered oryx; most of the south of the island is preserved for the ruling family's falconry
*head of state and government* Sheik Isa bin Sulman al-Khalifa from 1961
*political system* absolute emirate
*exports* oil, natural gas, aluminium, fish
*currency* Bahrain dinar
*population* (1990 est) 512,000 (two-thirds are nationals); growth rate 4.4% p.a.
*life expectancy* men 67, women 71
*languages* Arabic (official); Farsi, English, Urdu
*religion* 85% Muslim (Shi'ite 60%, Sunni 40%)
*literacy* men 79%, women 64% (1985 est)

**GDP** $3.5 bn (1987); $7,772 per head

*chronology*

**1861** Became British protectorate.
**1968** Britain announced its intention to withdraw its forces. Bahrain formed, with Qatar and the Trucial States, the Federation of Arab Emirates.
**1971** Qatar and the Trucial States withdrew from the federation and Bahrain became an independent state.
**1973** New constitution adopted, with an elected national assembly.
**1975** Prime minister resigned and national assembly dissolved. Emir and his family assumed virtually absolute power.
**1986** Gulf University established in Bahrain. A causeway was opened linking the island with Saudi Arabia.
**1988** Bahrain recognized Afghan rebel government.
**1991** Bahrain joined United Nations coalition that ousted Iraq from its occupation of Kuwait.

# Bangladesh

**Bangladesh**
People's Republic of
(*Gana Prajatantri Bangladesh*)
(formerly *East Pakistan*)

*area* 144,000 sq km/55,585 sq mi
*capital* Dhaka (formerly Dacca)
*towns* Chittagong, Khulna
*physical* flat delta of rivers Ganges (Padma) and Brahmaputra (Jamuna), the largest estuarine delta in the world; annual rainfall of 2,540 mm/100 in; some 75% of the land is less than 3 m/10 ft above sea level and vulnerable to flooding and cyclones; hilly in extreme SE and NE
*environment* deforestation on the slopes of the Himalayas increases the threat of flooding in the coastal lowlands, which are also subject to devastating monsoon storms. The building of India's Farakka Barrage has reduced the flow of the Ganges in Bangladesh and permitted salt water to intrude further inland. Increased salinity has destroyed fisheries, contaminated drinking water, and damaged forests
*head of state* Abdur Rahman Biswas from 1991
*head of government* Begum Khaleda Zia from 1991
*political system* emergent democratic republic
*exports* jute, tea, garments, fish products

*currency* taka
*population* (1991 est) 107,992,100; growth rate 2.17% p.a.; just over 1 million people live in small ethnic groups in the tropical Chittagong Hill Tracts, Mymensingh, and Sylhet districts
*life expectancy* men 50, women 52
*language* Bangla (Bengali)
*religions* Sunni Muslim 85%, Hindu 14%
*literacy* men 43%, women 22% (1985 est)
*GDP* $17.6 bn (1987); $172 per head
*chronology*
*1947* Formed into eastern province of Pakistan on partition of British India.
*1970* Half a million killed in flood.
*1971* Bangladesh emerged as independent nation, under leadership of Sheik Mujibur Rahman, after civil war.
*1975* Mujibur Rahman assassinated. Martial law imposed.
*1976–77* Maj-Gen Zia ur-Rahman assumed power.
*1978–79* Elections held and civilian rule restored.
*1981* Assassination of Maj-Gen Zia.
*1982* Lt-Gen Ershad assumed power in army coup. Martial law reimposed.
*1986* Elections held but disputed. Martial law ended.
*1987* State of emergency declared in response to opposition demonstrations.
*1988* Assembly elections boycotted by main opposition parties. State of emergency lifted. Islam made state religion. Monsoon floods left 30 million homeless and thousands dead.
*1989* Power devolved to Chittagong Hill Tracts to end 14-year conflict between local people and army-protected settlers.
*1990* Following mass antigovernment protests, President Ershad resigned; Shahabuddin Ahmad became interim president.
*1991* Feb: elections resulted in coalition government with Bangladesh Nationalist Party (BNP) dominant. April: cyclone killed around 139,000 and left up to 10 million homeless. Sept: parliamentary government restored; Abdur Rahman Biswas elected president.

# Barbados

*area* 430 sq km/166 sq mi
*capital* Bridgetown
*towns* Speightstown, Holetown, Oistins
*physical* most easterly island of the West Indies; surrounded by coral reefs; subject to hurricanes June–Nov
*head of state* Elizabeth II from 1966, represented by governor general Hugh Springer from 1984
*head of government* prime minister Erskine Lloyd Sandiford from 1987
*political system* constitutional monarchy
*exports* sugar, rum, electronic components, clothing, cement
*currency* Barbados dollar
*population* (1990 est) 260,000; growth rate 0.5% p.a.
*life expectancy* men 70, women 75
*languages* English and Bajan (Barbadian English dialect)
*religions* 70% Anglican, 9% Methodist, 4% Roman Catholic
*literacy* 99% (1984)
*GDP* $1.4 bn (1987); $5,449 per head

## *chronology*

**1627** Became British colony; developed as a sugar-plantation economy, initially on basis of slavery.
**1834** Slaves freed.
**1951** Universal adult suffrage introduced. Barbados Labour Party (BLP) won general election.
**1954** Ministerial government established.
**1961** Independence achieved from Britain. Democratic Labour Party (DLP), led by Errol Barrow, in power.
**1966** Barbados achieved full independence within Commonwealth. Barrow became the new nation's first prime minister.
**1972** Diplomatic relations with Cuba established.
**1976** BLP, led by Tom Adams, returned to power.
**1983** Barbados supported US invasion of Grenada.
**1985** Adams died; Bernard St John became prime minister.
**1986** DLP, led by Barrow, returned to power.
**1987** Barrow died; Erskine Lloyd Sandiford became prime minister.
**1989** New National Democratic Party (NDP) opposition formed.
**1991** DLP, under Erskine Sandiford, won general election.

## Belarus
Republic of

*area* 207,600 sq km/80,100 sq mi
*capital* Minsk (Mensk)
*towns* Gomel, Vitebsk, Mogilev, Bobruisk, Grodno, Brest
*physical* more than 25% forested; rivers W Dvina, Dnieper and its tributaries, including the Pripet and Beresina; the Pripet Marshes in the E; mild and damp climate
*environment* large areas contaminated by fallout from Chernobyl
*head of state* Stanislav Shushkevich from 1991
*head of government* Vyacheslav Kebich from 1990
*political system* emergent democracy
*products* peat, agricultural machinery, fertilizers, glass, textiles, leather, salt, electrical goods, meat, dairy produce
*currency* rouble and dukat
*population* (1990) 10,200,000 (77% Byelorussian 'Eastern Slavs', 13% Russian, 4% Polish, 1% Jewish)
*languages* Byelorussian, Russian
*religions* Roman Catholic, Russian Orthodox, with Baptist and Muslim minorities

*chronology*
**1918–19** Briefly independent from Russia.
**1937–41** More than 100,000 people were shot in mass executions ordered by Stalin.
**1941–44** Occupied by Nazi Germany.
**1945** Became a founding member of the United Nations.
**1986** April: fallout from the Chernobyl nuclear reactor in Ukraine contaminated a large area.
**1989** Byelorussian Popular Front established as well as a more extreme nationalist organization, the Tolaka group.
**1990** Sept: Byelorussian established as state language and republican sovereignty declared.
**1991** April: Minsk hit by nationalist-backed general strike, calling for disbandment of Communist Party (CP) workplace cells. Aug: declared independence from Soviet Union in wake of failed anti-Gorbachev coup; CP suspended. Sept: Shushkevich elected president. Dec: Commonwealth of Independent States (CIS) formed in Minsk; Belarus accorded diplomatic recognition by USA.
**1992** Jan: admitted into Conference on Security and Cooperation in Europe (CSCE). May: protocols signed with USA agreeing to honour START disarmament treaty.

## Belgium
Kingdom of
(French *Royaume de Belgique*, Flemish *Koninkrijk België*)

*area* 30,510 sq km/11,784 sq mi
*capital* Brussels
*towns* Ghent, Liège, Charleroi, Bruges, Antwerp, Ostend
*physical* fertile coastal plain in NW, central rolling hills rise eastwards, hills and forest in SE
*environment* a 1989 government report judged the drinking water in Flanders to be 'seriously substandard'
*head of state* King Baudouin from 1951
*head of government* Jean-Luc Dehaene from 1992
*political system* liberal democracy
*exports* iron, steel, textiles, manufactured goods, petrochemicals, plastics, vehicles, diamonds
*currency* Belgian franc
*population* (1990 est) 9,895,000 (comprising Flemings and Walloons); growth rate 0.1% p.a.
*life expectancy* men 72, women 78
*languages* in the N (Flanders) Flemish 55%; in the S (Wallonia) Walloon (a French dialect) 32%; bilingual 11%; German (E border)

0.6%; all are official
*religion* Roman Catholic 75%
*literacy* 98% (1984)
*GDP* $111 bn (1986); $9,230 per head
*chronology*
*1830* Belgium became an independent kingdom.
*1914* Invaded by Germany.
*1940* Again invaded by Germany.
*1948* Belgium became founding member of Benelux Customs Union.
*1949* Belgium became founding member of Council of Europe and NATO.
*1951* Leopold III abdicated in favour of his son Baudouin.
*1957* Belgium became founding member of the European Economic Community.
*1971* Steps towards regional autonomy taken.
*1972* German-speaking members included in the cabinet for the first time.
*1973* Linguistic parity achieved in government appointments.
*1974* Leo Tindemans became prime minister. Separate regional councils and ministerial committees established.
*1978* Wilfried Martens succeeded Tindemans as prime minister.
*1980* Open violence over language divisions. Regional assemblies for Flanders and Wallonia and a three-member executive for Brussels created.
*1981* Short-lived coalition led by Mark Eyskens was followed by the return of Martens.
*1987* Martens head of caretaker government after break-up of coalition.
*1988* Following a general election, Martens formed a new coalition between the Flemish Christian Social Party (CVP), French Socialist Party (PS), Flemish Socialist Party (SP), French Social Christian Party (PSC), and the Flemish People's Party (VU).
*1992* Martens-led coalition collapsed; Jean-Luc Dehaene formed a new CVP-led coalition. It was announced that a federal system would be introduced.

## Belize
(formerly *British Honduras*)

*area* 22,963 sq km/8,864 sq mi
*capital* Belmopan
*towns* Belize City, Dangriga, Punta Gorda, Orange Walk, Corozal
*physical* tropical swampy coastal plain, Maya Mountains in S; over 90% forested
*environment* since 1981 Belize has developed an extensive system of national parks and reserves to protect large areas of tropical forest, coastal mangrove, and offshore islands. Forestry has been replaced by agriculture and ecotourism, which are now the most important sectors of the economy; world's first jaguar reserve created 1986 in the Cockscomb Mountains
*head of state* Elizabeth II from 1981, represented by governor general
*head of government* George Price from 1989
*political system* constitutional monarchy
*exports* sugar, citrus fruits, rice, fish products, bananas
*currency* Belize dollar
*population* (1990 est) 180,400 (including Mayan minority in the interior); growth rate 2.5% p.a.
*life expectancy* (1988) 60 years

*languages* English (official); Spanish (widely spoken), native Creole dialects
*religions* Roman Catholic 60%, Protestant 35%
*literacy* 93% (1988)
*GDP* $247 million (1988); $1,220 per head
*chronology*
*1862* Belize became a British colony.
*1954* Constitution adopted, providing for limited internal self-government. General election won by George Price.
*1964* Self-government achieved from the UK (universal adult suffrage introduced).
*1965* Two-chamber national assembly introduced, with Price as prime minister.
*1970* Capital moved from Belize City to Belmopan.
*1973* British Honduras became Belize.
*1975* British troops sent to defend the disputed frontier with Guatemala.
*1977* Negotiations undertaken with Guatemala but no agreement reached.
*1980* United Nations called for full independence.
*1981* Full independence achieved. Price became prime minister.
*1984* Price defeated in general election. Manuel Esquivel formed the government. The UK reaffirmed its undertaking to defend the frontier.
*1989* Price and the People's United Party (PUP) won the general election.
*1991* Diplomatic relations with Guatemala established.

## Benin

People's Republic of
(*République
Populaire du Bénin*)

*area* 112,622 sq km/43,472 sq mi
*capital* Porto Novo (official), Cotonou (de facto)
*towns* Abomey, Natitingou, Parakou, Cotonou
*physical* flat to undulating terrain; hot and humid in S; semi-arid in N
*head of state and government* Nicéphore Soglo from 1991
*political system* socialist pluralist republic
*exports* cocoa, peanuts, cotton, palm oil, petroleum, cement
*currency* CFA franc
*population* (1990 est) 4,840,000; growth rate 3% p.a.
*life expectancy* men 42, women 46
*languages* French (official); Fon 47% and Yoruba 9% in south; six major tribal languages in north
*religions* animist 65%, Christian 17%, Muslim 13%
*literacy* men 37%, women 16% (1985 est)
*GDP* $1.6 bn (1987); $365 per head
*chronology*
*1851* Under French control.
*1958* Became self-governing dominion within the French Community.
*1960* Independence achieved from France.
*1960–72* Acute political instability, with switches from civilian to military rule.

*1972* Military regime established by General Mathieu Kerekou.
*1974* Kerekou announced that the country would follow a path of 'scientific socialism'.
*1975* Name of country changed from Dahomey to Benin.
*1977* Return to civilian rule under a new constitution.
*1980* Kerekou formally elected president by the national revolutionary assembly.
*1989* Marxist-Leninism dropped as official ideology. Strikes and protests against Kerekou's rule mounted; demonstrations banned and army deployed against protesters.
*1990* Referendum support for multiparty politics.
*1991* Multiparty elections held. Kerekou defeated in presidential elections by Nicéphore Soglo.

**Bhutan**
Kingdom of
(*Druk-yul*)

*area* 46,500 sq km/17,954 sq mi
*capital* Thimbu (Thimphu)
*towns* Paro, Punakha, Mongar
*physical* occupies southern slopes of the Himalayas; cut by valleys formed by tributaries of the Brahmaputra; thick forests in S
*head of state and government* Jigme Singye Wangchuk from 1972
*political system* absolute monarchy
*exports* timber, talc, fruit and vegetables, cement, distilled spirits, calcium carbide
*currency* ngultrum; also Indian currency
*population* (1990 est) 1,566,000; growth rate 2% p.a. (75% Ngalops and Sharchops, 25% Nepalese)
*life expectancy* men 44, women 43
*languages* Dzongkha (official, a Tibetan dialect), Sharchop, Bumthap, Nepali, and English
*religions* 75% Lamaistic Buddhist (state religion), 25% Hindu
*literacy* 5%
*GDP* $250 million (1987); $170 per head

## chronology

*1865* Trade treaty with Britain signed.
*1907* First hereditary monarch installed.
*1910* Anglo-Bhutanese Treaty signed.
*1949* Indo-Bhutan Treaty of Friendship signed.
*1952* King Jigme Dorji Wangchuk installed.
*1953* National assembly established.
*1959* 4,000 Tibetan refugees given asylum.
*1968* King established first cabinet.
*1972* King died and was succeeded by his son Jigme Singye Wangchuk.
*1979* Tibetan refugees told to take Bhutanese citizenship or leave; most stayed.
*1983* Bhutan became a founding member of the South Asian Regional Association for Cooperation (SAARC).
*1988* King imposed 'code of conduct' suppressing Nepalese customs.
*1990* Hundreds of people allegedly killed during prodemocracy demonstrations.

## Bolivia
Republic of
(*República de Bolivia*)

*area* 1,098,581 sq km/424,052 sq mi
*capital* La Paz (seat of government), Sucre (legal capital)
*towns* Santa Cruz, Cochabamba, Oruro, Potosí
*physical* high plateau (Altiplano) between mountain ridges (cordilleras); forest and lowlands (llano) in the E
*head of state and government* Jaime Paz Zamora from 1989
*political system* emergent democratic republic
*exports* tin, antimony (second largest world producer), other nonferrous metals, oil, gas, coffee, sugar, cotton
*currency* boliviano
*population* (1990 est) 6,730,000; (Quechua 25%, Aymara 17%, mestizo (mixed) 30%, European 14%); growth rate 2.7% p.a.
*life expectancy* men 51, women 54
*languages* Spanish, Aymara, Quechua (all official)
*religion* Roman Catholic 95% (state-recognized)
*literacy* men 84%, women 65% (1985 est)

**GDP** $4.2 bn (1987); $617 per head

*chronology*

*1825* Liberated from Spanish rule by Simón Bolívar; independence achieved (formerly known as Upper Peru).
*1952* Dr Víctor Paz Estenssoro elected president.
*1956* Dr Hernán Siles Zuazo became president.
*1960* Estenssoro returned to power.
*1964* Army coup led by the vice president, General René Barrientos.
*1966* Barrientos became president.
*1967* Uprising, led by 'Che' Guevara, put down with US help.
*1969* Barrientos killed in plane crash, replaced by Vice President Siles Salinas. Army coup deposed him.
*1970* Army coup put General Juan Torres González in power.
*1971* Torres replaced by Col Hugo Banzer Suárez.
*1973* Banzer promised a return to democratic government.
*1974* Attempted coup prompted Banzer to postpone elections and ban political and trade-union activity.
*1978* Elections declared invalid after allegations of fraud.
*1980* More inconclusive elections followed by another coup, led by General Luis García. Allegations of corruption and drug trafficking led to cancellation of US and EC aid.
*1981* García replaced by General Celso Torrelio Villa.
*1982* Torrelio resigned. Replaced by military junta led by General Guido Vildoso. Because of worsening economy, Vildoso asked congress to install a civilian administration. Dr Siles Zuazo chosen as president.
*1983* Economic aid from USA and Europe resumed.
*1984* New coalition government formed by Siles. Abduction of president by right-wing officers. The president undertook a five-day hunger strike as an example to the nation.
*1985* President Siles resigned. Election result inconclusive; Dr Paz Estenssoro, at the age of 77, chosen by congress as president.
*1989* Jaime Paz Zamora, Movement of the Revolutionary Left (MIR) elected president in power-sharing arrangement with Hugo Banzer Suárez, pledged to preserve free-market policies.
*1992* The new Solidarity Civil Union (UCS) party gained support.

## Bosnia-Herzegovina
Republic of

*area* 51,129 sq km/19,745 sq mi
*capital* Sarajevo
*towns* Banja Luka, Mostar, Prijedor, Tuzla, Zenica
*physical* barren, mountainous country
*population* (1990) 4,300,000 including 44% Muslims, 33% Serbs, 17% Croats; a complex patchwork of ethnically mixed communities
*head of state* Alija Izetbegović from 1990
*head of government* Mile Akmadzic from 1992
*political system* emergent democracy
*products* citrus fruits and vegetables; iron, steel, and leather goods; textiles
*language* Serbian variant of Serbo-Croatian
*religions* Sunni Muslim, Serbian Orthodox, Roman Catholic
*chronology*
*1918* Incorporated in the future Yugoslavia.
*1941* Occupied by Nazi Germany.
*1945* Became republic within Yugoslav Socialist Federation.
*1980* Upsurge in Islamic nationalism.

## Bosnia-Herzegovina

*1990* Ethnic violence erupted between Muslims and Serbs. Nov–Dec: communists defeated in multiparty elections; coalition formed by Serb, Muslim, and Croatian parties.

*1991* May: Serbia–Croatia conflict spread disorder into Bosnia-Herzegovina. Aug: Serbia revealed plans to annex the SE part of the republic. Sept: Serbian enclaves established by force. Oct: 'sovereignty' declared. Nov: plebiscite by Serbs favoured remaining within Yugoslavia; Serbs and Croats established autonomous communities.

*1992* Feb–March: Muslims and Croats voted overwhelmingly in favour of independence; referendum boycotted by Serbs. April: USA and EC recognized Bosnian independence. Ethnic hostilities escalated, with Serb forces occupying E and Croatian forces much of W; state of emergency declared; all-out civil war ensued. May: admitted to United Nations. June: Canadian–French UN forces drafted into Sarajevo to break three-month siege of city by Serbs. July: Canadian forces replaced by French, Egyptians, and Ukrainians. Official cease-fire broken intermittently by both sides; UN and EC mediators vainly sought truce. Fighting continued, with accusations of 'ethnic cleansing' being practised, particularly by Serbs. Oct: UN Security Council voted to create a war crimes commission and imposed ban on military flights over Bosnia-Herzegovina. First British troops deployed.

*1993* Jan: UN–EC peace plan, proposing to divide country into 10 autonomous, ethnically-controlled provinces, accepted in principle by Serbs and Croats but fighting continued. March: USA began airdrops of food and medical supplies.

# Botswana
Republic of

*area* 582,000 sq km/225,000 sq mi
*capital* Gaborone
*towns* Mahalpye, Serowe, Tutume, Francistown
*physical* desert in SW, plains in E, fertile lands and swamp in N
*environment* the Okavango Swamp is threatened by plans to develop the area for mining and agriculture
*head of state and government* Quett Ketamile Joni Masire from 1980
*political system* democratic republic
*exports* diamonds (third largest producer in world), copper, nickel, meat products, textiles
*currency* pula
*population* (1990 est) 1,218,000 (Bamangwato 80%, Bangwaketse 20%); growth rate 3.5% p.a.
*life expectancy* (1988) 59 years
*languages* English (official), Setswana (national)
*religions* Christian 50%, animist 50%
*literacy* (1988) 84%

***GDP*** $2.0 bn (1988); $1,611 per head

*chronology*

**1885** Became a British protectorate.
**1960** New constitution created a legislative council.
**1963** End of rule by High Commission.
**1965** Capital transferred from Mafeking to Gaborone. Internal self-government achieved. Sir Seretse Khama elected head of government.
**1966** Independence achieved from Britain. New constitution came into effect; name changed from Bechuanaland to Botswana; Seretse Khama elected president.
**1980** Seretse Khama died; succeeded by Vice President Quett Masire.
**1984** Masire re-elected.
**1985** South African raid on Gaborone.
**1987** Joint permanent commission with Mozambique established, to improve relations.
**1989** The Botswana Democratic Party (BDP) and Masire re-elected.

## Brazil
Federative Republic of
(*República Federativa do Brasil*)

*area* 8,511,965 sq km/3,285,618 sq mi
*capital* Brasília
*towns* São Paulo, Belo Horizonte, Curitiba, Fortaleza, Rio de Janeiro, Recife, Salvador
*physical* the densely forested Amazon basin covers the northern half of the country with a network of rivers; the south is fertile
*environment* Brazil has one-third of the world's tropical rainforest. It contains 55,000 species of flowering plants (the greatest variety in the world) and 20% of all the world's bird species. During the 1980s at least 7% of the Amazon rainforest was destroyed by settlers who cleared the land for cultivation and grazing
*head of state and government* Itamar Franco from 1992
*political system* emergent democratic federal republic
*exports* coffee, sugar, soya beans, cotton, textiles, timber, motor vehicles, iron, chrome, manganese, tungsten and other ores; the world's sixth largest arms exporter

***currency*** cruzado (introduced 1986); inflation 1990 was 1,795%
***population*** (1990 est) 153,770,000 (including 200,000 Indians, mostly living on reservations); growth rate 2.2% p.a.
***life expectancy*** men 61, women 66
***languages*** Portuguese (official); 120 Indian languages
***religions*** Roman Catholic 89%; Indian faiths
***literacy*** men 79%, women 76% (1985 est)
***GDP*** \$352 bn (1988); \$2,434 per head
***chronology***
*1822* Independence achieved from Portugal.
*1889* Monarchy abolished and republic established.
*1891* Constitution for a federal state adopted.
*1960* Capital moved to Brasília from Rio de Janeiro.
*1961* João Goulart became president.
*1964* Bloodless coup made General Castelo Branco president; he assumed dictatorial powers, abolishing free political parties.
*1967* New constitution adopted. Branco succeeded by Marshal da Costa e Silva.
*1969* Da Costa e Silva resigned and a military junta took over.
*1974* General Ernesto Geisel became president.
*1978* General Baptista de Figueiredo became president.
*1979* Political parties legalized again.
*1984* Mass calls for a return to fully democratic government.
*1985* Tancredo Neves became first civilian president in 21 years. Neves died and was succeeded by the vice president, José Sarney.
*1988* New constitution approved, transferring power from the president to the congress. Measures announced to halt large-scale burning of Amazonian rainforest for cattle grazing.
*1989* Forest Protection Service and Ministry for Land Reform abolished. International concern over how much of the Amazon has been burned. Fernando Collor, National Reconstruction Party (PRN), elected president, pledging free-market economic policies.
*1990* Government won the general election offset by mass abstentions.
*1992* June: Earth Summit, global conference on the environment, held in Rio de Janeiro. Sept: Collor charged with corruption and stripped of his powers. Replaced by Vice President Itamar Franco.

# Brunei

Islamic Sultanate of
(*Negara Brunei Darussalam*)

***area*** 5,765 sq km/2,225 sq mi
***capital*** Bandar Seri Begawan
***towns*** Tutong, Seria, Kuala Belait
***physical*** flat coastal plain with hilly lowland in W and mountains in E; 75% of the area is forested; the Limbang valley splits Brunei in two, and its cession to Sarawak 1890 is disputed by Brunei
***head of state and of government*** HM Muda Hassanal Bolkiah Mu'izzaddin Waddaulah, Sultan of Brunei, from 1968
***political system*** absolute monarchy
***exports*** liquefied natural gas (world's largest producer) and oil, both expected to be exhausted by the year 2000
***currency*** Brunei dollar
***population*** (1990 est) 372,000 (65% Malay, 20% Chinese – few Chinese granted citizenship); growth rate 12% p.a.
***life expectancy*** 74 years
***languages*** Malay (official), Chinese (Hokkien), English
***religion*** 60% Muslim (official)

*literacy* 95%
*GDP* $3.4 bn (1985); $20,000 per head
*chronology*
*1888* Brunei became a British protectorate.
*1941–45* Occupied by Japan.
*1959* Written constitution made Britain responsible for defence and external affairs.
*1962* Sultan began rule by decree.
*1963* Proposal to join Malaysia abandoned.
*1967* Sultan abdicated in favour of his son, Hassanal Bolkiah.
*1971* Brunei given internal self-government.
*1975* United Nations resolution called for independence for Brunei.
*1984* Independence achieved from Britain, with Britain maintaining a small force to protect the oil- and gasfields.
*1985* A 'loyal and reliable' political party, the Brunei National Democratic Party (BNDP), legalized.
*1986* Death of former sultan, Sir Omar. Formation of multiethnic Brunei National United Party (BNUP).
*1988* BNDP banned.

# Bulgaria
Republic of
(*Republika Bulgaria*)

*area* 110,912 sq km/42,812 sq mi
*capital* Sofia
*towns* Plovdiv, Ruse, Burgas, Varna
*physical* lowland plains in N and SE separated by mountains that cover three-quarters of the country
*environment* pollution has virtually eliminated all species of fish once caught in the Black Sea. Vehicle-exhaust emissions in Sofia have led to dust concentrations more than twice the medically accepted level
*head of state* Zhelyu Zhelev from 1990
*head of government* Lyuben Berov from 1992
*political system* emergent democratic republic
*exports* textiles, leather, chemicals, nonferrous metals, timber, machinery, tobacco, cigarettes (world's largest exporter)
*currency* lev
*population* (1990 est) 8,978,000 (including 900,000–1,500,000 ethnic Turks, concentrated in S and NE); growth rate 0.1% p.a.
*life expectancy* men 69, women 74
*languages* Bulgarian, Turkish

***religions*** Eastern Orthodox Christian 90%, Sunni Muslim 10%
***literacy*** 98%
***GDP*** $25.4 bn (1987); $2,836 per head
*chronology*
***1908*** Bulgaria became a kingdom independent of Turkish rule.
***1944*** Soviet invasion of German-occupied Bulgaria.
***1946*** Monarchy abolished and communist-dominated people's republic proclaimed.
***1947*** Soviet-style constitution adopted.
***1949*** Death of Georgi Dimitrov, the communist government leader.
***1954*** Election of Todor Zhivkov as Communist Party general secretary; made nation a loyal satellite of USSR.
***1971*** Constitution modified; Zhivkov elected president.
***1985–89*** Large administrative and personnel changes made haphazardly under Soviet stimulus.
***1987*** New electoral law introduced multicandidate elections.
***1989*** Programme of 'Bulgarianization' resulted in mass exodus of Turks to Turkey. Nov: Zhivkov ousted by Petar Mladenov. Dec: opposition parties allowed to form.
***1990*** April: Bulgarian Communist Party (BCP) renamed Bulgarian Socialist Party (BSP). Aug: Dr Zhelyu Zhelev elected president. Nov: government headed by Andrei Lukanov resigned, replaced Dec by coalition led by Dimitur Popov.
***1991*** July: new constitution adopted. Oct: Union of Democratic Forces (UDF) beat BSP in general election by narrow margin; formation of first noncommunist, UDF-minority government under Filip Dimitrov.
***1992*** Zhelev became Bulgaria's first directly elected president. Relations with West greatly improved. Dimitrov resigned after vote of no confidence; replaced by Lyuben Berov.

# Burkina Faso

The People's Democratic Republic of (formerly *Upper Volta*)

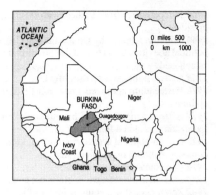

*area* 274,122 sq km/105,811 sq mi
*capital* Ouagadougou
*towns* Bobo-Dioulasso, Koudougou
*physical* landlocked plateau with hills in W and SE; headwaters of the river Volta; semi-arid in N, forest and farmland in S
*environment* tropical savanna subject to overgrazing and deforestation
*head of state and government* Blaise Compaoré from 1987
*political system* transitional
*exports* cotton, karite nuts, groundnuts, livestock, hides, skins, sesame, cereals
*currency* CFA franc
*population* (1990 est) 8,941,000; growth rate 2.4% p.a.
*life expectancy* men 44, women 47
*languages* French (official); about 50 native Sudanic languages spoken by 90% of population
*religions* animist 53%, Sunni Muslim 36%, Roman Catholic 11%
*literacy* men 21%, women 6% (1985 est)
*GDP* $1.6 bn (1987); $188 per head
*chronology*
*1958* Became a self-governing republic within the French Community.

*1960* Independence from France, with Maurice Yaméogo as the first president.
*1966* Military coup led by Col Lamizana. Constitution suspended, political activities banned, and a supreme council of the armed forces established.
*1969* Ban on political activities lifted.
*1970* Referendum approved a new constitution leading to a return to civilian rule.
*1974* After experimenting with a mixture of military and civilian rule, Lamizana reassumed full power.
*1977* Ban on political activities removed. Referendum approved a new constitution based on civilian rule.
*1978* Lamizana elected president.
*1980* Lamizana overthrown in bloodless coup led by Col Zerbo.
*1982* Zerbo ousted in a coup by junior officers. Major Ouédraogo became president and Thomas Sankara prime minister.
*1983* Sankara seized complete power.
*1984* Upper Volta renamed Burkina Faso, 'land of upright men'.
*1987* Sankara killed in coup led by Blaise Compaoré.
*1989* New government party Organization for Popular Democracy–Workers' Party (ODP–MT) formed by merger of other pro-government parties. Coup against Compaoré foiled.
*1991* New constitution approved. Compaoré re-elected president.
*1992* Multiparty elections won by FP–Popular Front.

# Burundi
Republic of
(*Republika y'Uburundi*)

*area* 27,834 sq km/10,744 sq mi
*capital* Bujumbura
*towns* Gitega, Bururi, Ngozi, Muyinga
*physical* landlocked grassy highland straddling watershed of Nile and Congo
*head of state and government* Pierre Buyoya from 1987
*political system* one-party military republic
*exports* coffee, cotton, tea, nickel, hides, livestock, cigarettes, beer, soft drinks; there are 500 million tonnes of peat reserves in the basin of the Akanyaru River
*currency* Burundi franc
*population* (1990 est) 5,647,000 (of whom 15% are the Nilotic Tutsi, still holding most of the land and political power, 1% are Pygmy Twa, and the remainder Bantu Hutu); growth rate 2.8% p.a.
*life expectancy* men 45, women 48
*languages* Kirundi (a Bantu language) and French (both official), Kiswahili
*religions* Roman Catholic 62%, Protestant 5%, Muslim 1%, animist 32%
*literacy* men 43%, women 26% (1985)
*GDP* $1.1 bn (1987); $230 per head

## chronology

**1962** Separated from Ruanda-Urundi, as Burundi, and given independence as a monarchy under King Mwambutsa IV.
**1966** King deposed by his son Charles, who became Ntare V; he was in turn deposed by his prime minister, Capt Michel Micombero, who declared Burundi a republic.
**1972** Ntare V killed, allegedly by the Hutu ethnic group. Massacres of 150,000 Hutus by the rival Tutsi ethnic group, of which Micombero was a member.
**1973** Micombero made president and prime minister.
**1974** Union for National Progress (UPRON) declared the only legal political party, with the president as its secretary general.
**1976** Army coup deposed Micombero. Col Jean-Baptiste Bagaza appointed president by the Supreme Revolutionary Council.
**1981** New constitution adopted, providing for a national assembly.
**1984** Bagaza elected president as sole candidate.
**1987** Bagaza deposed in coup Sept. Maj Pierre Buyoya headed new Military Council for National Redemption.
**1988** Some 24,000 majority Hutus killed by Tutsis.
**1992** New constitution approved.

## Cambodia

State of (formerly *Khmer Republic* 1970–76, *Democratic Kampuchea* 1976–79, *People's Republic of Kampuchea* 1979–89)

*area* 181,035 sq km/69,880 sq mi
*capital* Phnom Penh
*towns* Battambang, Kompong Som
*physical* mostly flat forested plains with mountains in SW and N; Mekong River runs N–S
*head of state* Prince Norodom Sihanouk from 1991
*head of government* Hun Sen from 1985
*political system* transitional
*exports* rubber, rice, pepper, wood, cattle
*currency* Cambodian riel
*population* (1990 est) 6,993,000; growth rate 2.2% p.a.
*life expectancy* men 42, women 45
*languages* Khmer (official), French
*religion* Theravāda Buddhist 95%
*literacy* men 78%, women 39% (1980 est)
*GDP* $592 mn (1987); $83 per head

*chronology*
**1863–1941** French protectorate.
**1941–45** Occupied by Japan.
**1946** Recaptured by France.
**1953** Independence achieved from France.
**1970** Prince Sihanouk overthrown by US-backed Lon Nol.
**1975** Lon Nol overthrown by Khmer Rouge.
**1976–78** Khmer Rouge introduced an extreme communist programme, forcing urban groups into rural areas and bringing about over 2.5 million deaths from famine, disease, and maltreatment.
**1978–79** Vietnamese invasion and installation of Heng Samrin government.
**1982** The three main anti-Vietnamese resistance groups formed an alliance under Prince Sihanouk.
**1987** Vietnamese troop withdrawal began.
**1989** Sept: completion of Vietnamese withdrawal. Nov: United Nations peace proposal rejected by Phnom Penh government.
**1991** Oct: Peace agreement signed in Paris, providing for a UN Transitional Authority in Cambodia (UNTAC) to administer country in transition period in conjunction with all-party Supreme National Council; communism abandoned. Nov: Sihanouk returned as head of state.
**1992** Political prisoners released; freedom of speech and party formation restored. Oct: Khmer Rouge refused to disarm in accordance with peace process. Dec: UN Security Council voted to impose limited trade embargo on area of country controlled by Khmer Rouge guerrillas.
**1993** General election set for May.

## Cameroon
Republic of
(*République du Cameroun*)

*area* 475,440 sq km/183,638 sq mi
*capital* Yaoundé
*towns* Douala, Nkongsamba, Garova
*physical* desert in far north in the Lake Chad basin, mountains in W, dry savanna plateau in the intermediate area, and dense tropical rainforest in S
*environment* the Korup National Park preserves 1,300 sq km/ 500 sq mi of Africa's fast-disappearing tropical rainforest
*head of state and of government* Paul Biya from 1982
*political system* emergent democratic republic
*exports* cocoa, coffee, bananas, cotton, timber, rubber, groundnuts, gold, aluminium, crude oil
*currency* CFA franc
*population* (1990 est) 11,109,000; growth rate 2.7% p.a.
*life expectancy* men 49, women 53
*languages* French and English in pidgin variations (official); there has been some discontent with the emphasis on French – there are 163 indigenous peoples with their own African languages
*religions* Roman Catholic 35%, animist 25%, Muslim 22%, Protestant 18%
*literacy* men 68%, women 45% (1985 est)

**GDP** $12.7 bn (1987); $1,170 per head

*chronology*

*1884* Treaty signed establishing German rule.
*1916* Captured by Allied forces in World War I.
*1922* Divided between Britain and France.
*1946* French Cameroon and British Cameroons made UN trust territories.
*1960* French Cameroon became the independent Republic of Cameroon. Ahmadou Ahidjo elected president.
*1961* Northern part of British Cameroon merged with Nigeria and southern part joined the Republic of Cameroon to become the Federal Republic of Cameroon.
*1966* One-party regime introduced.
*1972* New constitution made Cameroon a unitary state, the United Republic of Cameroon.
*1973* New national assembly elected.
*1982* Ahidjo resigned and was succeeded by Paul Biya.
*1983* Biya began to remove his predecessor's supporters; accused by Ahidjo of trying to create a police state. Ahidjo went into exile in France.
*1984* Biya re-elected; defeated a plot to overthrow him. Country's name changed to Republic of Cameroon.
*1988* Biya re-elected.
*1990* Widespread public disorder. Biya granted amnesty to political prisoners.
*1991* Constitutional changes made.
*1992* Ruling Democratic Assembly of the Cameroon People (RDPC) won in first multiparty elections in 28 years. Biya's presidential victory challenged by opposition.

# Canada

*area* 9,970,610 sq km/3,849,674 sq mi
*capital* Ottawa
*towns* Toronto, Montréal, Vancouver, Edmonton, Calgary, Winnipeg, Quebec
*physical* mountains in W, with low-lying plains in interior and rolling hills in E. Climate varies from temperate in S to arctic in N
*environment* sugar maples are dying in E Canada as a result of increasing soil acidification
*head of state* Elizabeth II from 1952, represented by governor general
*head of government* Brian Mulroney from 1984
*political system* federal constitutional monarchy
*exports* wheat, timber, pulp, newsprint, salmon, furs (ranched fox and mink exceed the value of wild furs), oil, natural gas, aluminium, asbestos, coal, copper, iron, zinc, nickel and uranium (world's largest producer), motor vehicles and parts, fertilizers, chemicals
*currency* Canadian dollar

***population*** (1990 est) 26,527,000—including 300,000 North American Indians, of whom 75% live on more than 2,000 reservations in Ontario and the four western provinces; some 300,000 Métis (people of mixed race) and 19,000 Inuit of whom 75% live in the Northwest Territories. Growth rate 1.1% p.a.
***life expectancy*** men 72, women 79
***languages*** English, French (both official); there are also North American Indian languages and the Inuit Inuktitut
***religion*** Roman Catholic 46%, Protestant 35%
***literacy*** 99%
***GDP*** $412 bn (1987); $15,910 per head
*chronology*
***1867*** Dominion of Canada founded.
***1949*** Newfoundland joined Canada.
***1957*** Progressive Conservatives returned to power.
***1963*** Liberals elected under Lester Pearson.
***1968*** Pearson succeeded by Pierre Trudeau.
***1979*** Joe Clark, leader of the Progressive Conservatives, formed a minority government; defeated on budget proposals.
***1980*** Liberals under Trudeau returned with a large majority. Québec referendum rejected demand for independence.
***1982*** Canada Act removed Britain's last legal control over Canadian affairs; 'patriation' of Canada's constitution.
***1983*** Brian Mulroney became leader of the Progressive Conservatives.
***1984*** Trudeau retired and was succeeded as Liberal leader and prime minister by John Turner. Progressive Conservatives won the federal election with a large majority, and Mulroney became prime minister.
***1988*** Conservatives re-elected with reduced majority on platform of free trade with the USA.
***1990*** Collapse of Meech Lake accord. Canada joined the coalition opposing Iraq's invasion of Kuwait.
***1991*** Constitutional reform package proposed.
***1992*** Self-governing homeland for Inuit approved. Constitutional reform package, the Charlottetown agreement, rejected in referendum.
***1993*** Feb: Mulroney resigned leadership of Conservative Party but remained prime minister until a successor was appointed.

## Cape Verde
Republic of
(*República de Cabo Verde*)

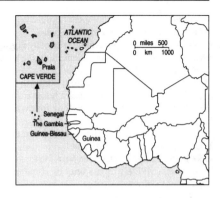

*area* 4,033 sq km/1,557 sq mi
*capital* Praia
*towns* Mindelo, Sal-Rei, Porto Novo
*physical* archipelago of ten volcanic islands 565 km/350 mi W of Senegal; the windward (Barlavento) group includes Santo Antão, São Vicente, Santa Luzia, São Nicolau, Sal, and Boa Vista; the leeward (Sotovento) group comprises Maio, São Tiago, Fogo, and Brava; all but Santa Luzia are inhabited
*head of state* Mascarenhas Monteiro from 1991
*head of government* Carlos Viega from 1991
*political system* socialist pluralist state
*exports* bananas, salt, fish
*currency* Cape Verde escudo
*population* (1990 est) 375,000 (including 100,000 Angolan refugees); growth rate 1.9% p.a.
*life expectancy* men 57, women 61
*language* Creole dialect of Portuguese
*religion* Roman Catholic 80%
*literacy* men 61%, women 39% (1985)
*GDP* $158 million (1987); $454 per head

*chronology*
*15th century* First settled by Portuguese.
*1951–74* Ruled as an overseas territory by Portugal.
*1974* Moved towards independence through a transitional Portuguese–Cape Verde government.
*1975* Independence achieved from Portugal. National people's assembly elected. Aristides Pereira became the first president.
*1980* Constitution adopted providing for eventual union with Guinea-Bissau.
*1981* Union with Guinea-Bissau abandoned and the constitution amended; became one-party state.
*1991* First multiparty elections held. New party, Movement for Democracy (MPD), won majority in assembly. Pereira replaced by Mascarenhas Monteiro.

# Central African Republic
(*République Centrafricaine*)

*area* 622,436 sq km/240,260 sq mi
*capital* Bangui
*towns* Berbérati, Bouar, Bossangoa
*physical* landlocked flat plateau, with rivers flowing N and S, and hills in NE and SW; dry in N, rainforest in SW
*environment* an estimated 87% of the urban population is without access to safe drinking water
*head of state and government* André Kolingba from 1981
*exports* diamonds, uranium, coffee, cotton, hardwood timber, tobacco
*currency* CFA franc
*population* (1990 est) 2,879,000 (more than 80 ethnic groups); growth rate 2.3% p.a.
*life expectancy* men 41, women 45
*languages* Sangho (national), French (official), Arabic, Hunsa, and Swahili
*religions*  Protestant 25%; Roman Catholic 25%;  Muslim 10%; animist 10%
*literacy* men 53%, women 29% (1985 est)
*GDP* $1 bn (1987); $374 per head

## Central African Republic

*chronology*

**1960** Central African Republic achieved independence from France; David Dacko elected president.
**1962** The republic made a one-party state.
**1965** Dacko ousted in military coup led by Col Bokassa.
**1966** Constitution rescinded and national assembly dissolved.
**1972** Bokassa declared himself president for life.
**1977** Bokassa made himself emperor of the Central African Empire.
**1979** Bokassa deposed by Dacko following violent repressive measures by the self-styled emperor, who went into exile.
**1981** Dacko deposed in a bloodless coup, led by General André Kolingba, and an all-military government established.
**1983** Clandestine opposition movement formed.
**1984** Amnesty for all political party leaders announced. President Mitterrand of France paid a state visit.
**1985** New constitution promised, with some civilians in the government.
**1986** Bokassa returned from France, expecting to return to power; he was imprisoned and his trial started. General Kolingba re-elected. New constitution approved by referendum.
**1988** Bokassa found guilty and received death sentence, later commuted to life imprisonment.
**1991** Government announced that a national conference would be held in response to demands for a return to democracy.
**1992** Abortive debate held on political reform; multiparty elections promised but then postponed.

## Chad
Republic of
(*République du Tchad*)

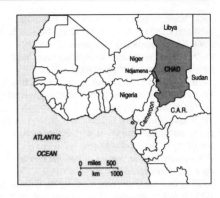

*area* 1,284,000 sq km/495,624 sq mi
*capital* Ndjamena (formerly Fort Lamy)
*towns* Sarh, Moundou, Abéché
*physical* landlocked state with mountains and part of Sahara Desert in N; moist savanna in S; rivers in S flow NW to Lake Chad
*head of state and government* Idriss Deby from 1990
*political system* emergent democratic republic
*exports* cotton, meat, livestock, hides, skins
*currency* CFA franc
*population* (1990 est) 5,064,000; growth rate 2.3% p.a. Nomadic tribes move N–S seasonally in search of water
*life expectancy* men 42, women 45
*languages* French, Arabic (both official), over 100 African languages are spoken
*religions* Muslim 44% (N), Christian 33%, animist 23% (S)
*literacy* men 40%, women 11% (1985 est)
*GDP* $980 million (1986); $186 per head
*chronology*
*1960* Independence achieved from France, with François Tombalbaye as president.

*1963* Violent opposition in the Muslim north, led by the Chadian National Liberation Front (Frolinat), backed by Libya.
*1968* Revolt quelled with France's help.
*1975* Tombalbaye killed in military coup led by Félix Malloum. Frolinat continued its resistance.
*1978* Malloum tried to find a political solution by bringing the former Frolinat leader Hissène Habré into his government but they were unable to work together.
*1979* Malloum forced to leave the country; an interim government was set up under General Goukouni. Habré continued his opposition with his Army of the North (FAN).
*1981* Habré now in control of half the country. Goukouni fled and set up a 'government in exile'.
*1983* Habré's regime recognized by the Organization for African Unity (OAU), but in the north Goukouni's supporters, with Libya's help, fought on. Eventually a cease-fire was agreed, with latitude 16°N dividing the country.
*1984* Libya and France agreed to a withdrawal of forces.
*1985* Fighting between Libyan-backed and French-backed forces intensified.
*1987* Chad, France, and Libya agreed on cease-fire proposed by OAU.
*1988* Full diplomatic relations with Libya restored.
*1989* Libyan troop movements reported on border; Habré re-elected, amended constitution.
*1990* President Habré ousted in coup led by Idriss Deby. New constitution adopted.
*1991* Several anti-government coups foiled.
*1992* Anti-government coup foiled. Two new opposition parties approved.

# Chile
Republic of
(*República de Chile*)

*area* 756,950 sq km/292,257 sq mi
*capital* Santiago
*towns* Concepción, Viña del Mar, Temuco; ports Valparaíso, Antofagasta, Arica, Iquique, Punta Arenas
*physical* Andes mountains along E border, Atacama Desert in N, fertile central valley, grazing land and forest in S
*territories* Easter Island, Juan Fernández Islands, part of Tierra del Fuego, claim to part of Antarctica
*head of state and government* Patricio Aylwin from 1990
*political system* emergent democratic republic
*exports* copper (world's leading producer), iron, molybdenum (world's second largest producer), nitrate, pulp and paper, fishmeal, fruit
*currency* peso
*population* (1990 est) 13,000,000 (the majority are of European origin or are mestizos, of mixed American Indian and Spanish descent); growth rate 1.6% p.a.

*life expectancy* men 64, women 73
*language* Spanish
*religion* Roman Catholic 89%
*literacy* 94% (1988)
*GDP* $18.9 bn (1987); $6,512 per head
*chronology*
**1818** Achieved independence from Spain.
**1964** Christian Democratic Party (PDC) formed government under Eduardo Frei.
**1970** Dr Salvador Allende became the first democratically elected Marxist president; he embarked on an extensive programme of nationalization and social reform.
**1973** Government overthrown by the CIA-backed military, led by General Augusto Pinochet. Allende killed. Policy of repression began during which all opposition was put down and political activity banned.
**1983** Growing opposition to the regime from all sides, with outbreaks of violence.
**1988** Referendum on whether Pinochet should serve a further term resulted in a clear 'No' vote.
**1989** President Pinochet agreed to constitutional changes to allow pluralist politics. Patricio Aylwin (PDC) elected president (his term would begin 1990); Pinochet remained as army commander in chief.
**1990** Aylwin reached accord on end to military junta government. Pinochet censured by president.
**1992** Future US–Chilean free-trade agreement announced.

# China
People's Republic of
(*Zhonghua Renmin Gonghe Guo*)

*area* 9,596,960 sq km/3,599,975 sq mi
*capital* Beijing (Peking)
*towns* Chongqing (Chungking), Shenyang (Mukden), Nanjing (Nanking), Harbin, Tianjin (Tientsin), Shanghai, Guangzhou (Canton)
*physical* two-thirds of China is mountains or desert (N and W); the low-lying E is irrigated by rivers Huang He (Yellow River), Chang Jiang (Yangtze-Kiang), Xi Jiang (Si Kiang)
*head of state* Yang Shangkun from 1988
*head of government* Li Peng from 1987
*political system* communist republic
*exports* tea, livestock and animal products, silk, cotton, oil, minerals (China is the world's largest producer of tungsten and antimony), chemicals, light industrial goods
*currency* yuan
*population* (1990 est) 1,130,065,000 (the majority are Han or ethnic Chinese; the 67 million of other ethnic groups, including Tibetan, Uigur, and Zhuang, live in border areas). The number of people of Chinese origin outside China, Taiwan, and Hong Kong is estimated at

15–24 million. Growth rate 1.2% p.a.
*life expectancy* men 67, women 69
*languages* Chinese, including Mandarin (official), Cantonese, and other dialects
*religions* officially atheist, but traditionally Taoist, Confucianist, and Buddhist; Muslim 13 million; Catholic 3–6 million (divided between the 'patriotic' church established 1958 and the 'loyal' church subject to Rome); Protestant 3 million
*literacy* men 82%, women 66% (1985 est)
*GDP* $293.4 bn (1987); $274 per head
*chronology*
*1949* People's Republic of China proclaimed by Mao Zedong.
*1954* Soviet-style constitution adopted.
*1956–57* Hundred Flowers Movement encouraged criticism of the government.
*1958–60* Great Leap Forward commune experiment to achieve 'true communism'.
*1960* Withdrawal of Soviet technical advisers.
*1962* Sino-Indian border war.
*1962–65* Economic recovery programme under Liu Shaoqi; Maoist 'socialist education movement' rectification campaign.
*1966–69* Great Proletarian Cultural Revolution; Liu Shaoqi overthrown.
*1969* Ussuri River border clashes with USSR.
*1970–76* Reconstruction under Mao and Zhou Enlai.
*1971* Entry into United Nations.
*1972* US president Nixon visited Beijing.
*1975* New state constitution. Unveiling of Zhou's 'Four Modernizations' programme.
*1976* Deaths of Zhou Enlai and Mao Zedong; appointment of Hua Guofeng as prime minister and Communist Party chair. Vice Premier Deng Xiaoping in hiding. Gang of Four arrested.
*1977* Rehabilitation of Deng Xiaoping.
*1979* Economic reforms introduced. Diplomatic relations opened with USA. Punitive invasion of Vietnam.
*1980* Zhao Ziyang appointed prime minister.

*1981* Hu Yaobang succeeded Hua Guofeng as party chair. Imprisonment of Gang of Four.
*1982* New state constitution adopted.
*1984* 'Enterprise management' reforms for industrial sector.
*1986* Student prodemocracy demonstrations.
*1987* Hu was replaced as party leader by Zhao, with Li Peng as prime minister. Deng left Politburo but remained influential.
*1988* Yang Shangkun replaced Li Xiannian as state president. Economic reforms encountered increasing problems; inflation rocketed.
*1989* Over 2,000 killed in prodemocracy student demonstrations in Tiananmen Square; international sanctions imposed.
*1991* March: European Community and Japanese sanctions lifted. May: normal relations with USSR resumed. Sept: UK prime minister John Major visited Beijing. Nov: relations with Vietnam normalized.
*1992* China promised to sign 1968 Nuclear Non-Proliferation Treaty. Historic visit by Japan's emperor.
*1993* Jiang Zemin, Chinese Communist Party general secretary, set to replace Yang Shangkun as president.

# Colombia

**Colombia**
Republic of
(*República de Colombia*)

*area* 1,141,748 sq km/440,715 sq mi
*capital* Bogotá
*towns* Medellín, Cali, Bucaramanga, Barranquilla, Cartagena,
*physical* the Andes mountains run N–S; flat coastland in W and plains (llanos) in E; Magdalena River runs N to Caribbean Sea; includes islands of Providencia, San Andrés, and Mapelo
*head of state and government* Cesar Gaviria Trujillo from 1990
*political system* emergent democratic republic
*exports* emeralds (world's largest producer), coffee (world's second largest producer), cocaine (country's largest export), bananas, cotton, meat, sugar, oil, skins, hides, tobacco
*currency* peso
*population* (1990 est) 32,598,800 (mestizo 68%, white 20%, Amerindian 1%); growth rate 2.2% p.a.
*life expectancy* men 61, women 66; Indians 34
*language* Spanish
*religion* Roman Catholic 95%
*literacy* men 89%, women 87% (1987); Indians 40%
*GDP* $31.9 bn (1987); $1,074 per head

*chronology*
**1886** Full independence achieved from Spain. Conservatives in power.
**1930** Liberals in power.
**1946** Conservatives in power.
**1948** Left-wing mayor of Bogotá assassinated; widespread outcry.
**1949** Start of civil war, 'La Violencia', during which over 250,000 people died.
**1957** Hoping to halt the violence, Conservatives and Liberals agreed to form a National Front, sharing the presidency.
**1970** National Popular Alliance (ANAPO) formed as a left-wing opposition to the National Front.
**1974** National Front accord temporarily ended.
**1975** Civil unrest because of disillusionment with the government.
**1978** Liberals, under Julio Turbay, revived the accord and began an intensive fight against drug dealers.
**1982** Liberals maintained their control of congress but lost the presidency. The Conservative president, Belisario Betancur, granted guerrillas an amnesty and freed political prisoners.
**1984** Minister of justice assassinated by drug dealers; campaign against them stepped up.
**1986** Virgilio Barco Vargas, Liberal, elected president by record margin.
**1989** Drug cartel assassinated leading presidential candidate; Vargas declared antidrug war; bombing campaign by drug lords killed hundreds; police killed José Rodríguez Gacha, one of the most wanted cartel leaders.
**1990** Cesar Gaviria Trujillo elected president. Liberals maintained control of congress.
**1991** New constitution prohibited extradition of Colombians wanted for trial in other countries; several leading drug traffickers arrested. Oct: Liberal Party won general election.
**1992** One of leading drug barons, Pablo Escobar, escaped from prison.
**1993** Escobar continued to defy government.

## Comoros

Federal Islamic Republic of
(*Jumhurīyat al-Qumur al-Itthādīyah al-Islāmīyah*)

*area* 1,862 sq km/719 sq mi
*capital* Moroni
*towns* Mutsamudu, Domoni, Fomboni
*physical* comprises the volcanic islands of Njazídja, Nzwani, and Mwali (formerly Grande Comore, Anjouan, Moheli); at N end of Mozambique Channel
*head of state* Said Mohammad Djohar (interim administration) from 1989
*head of government* Halidi Abderamane Ibrahim from 1993
*political system* authoritarian nationalism
*exports* copra, vanilla, cocoa, sisal, coffee, cloves, essential oils
*currency* CFA franc
*population* (1990 est) 459,000; growth rate 3.1% p.a.
*life expectancy* men 48, women 52
*languages* Arabic (official), Comorian (Swahili and Arabic dialect), Makua, French
*religions* Muslim (official) 86%, Roman Catholic 14%

*literacy* 15%
*GDP* $198 million (1987); $468 per head
*chronology*
*1975* Independence achieved from France, but island of Mayotte remained part of France. Ahmed Abdallah elected president. The Comoros joined the United Nations.
*1976* Abdallah overthrown by Ali Soilih.
*1978* Soilih killed by mercenaries working for Abdallah. Islamic republic proclaimed and Abdallah elected president.
*1979* The Comoros became a one-party state; powers of the federal government increased.
*1985* Constitution amended to make Abdallah head of government as well as head of state.
*1989* Abdallah killed by French mercenaries who took control of government; under French and South African pressure, mercenaries left Comoros, turning authority over to French administration and interim president Said Mohammad Djohar.
*1990* Antigovernment coup foiled.
*1992* Third transitional government appointed. Antigovernment coup foiled.
*1993* Jan: general election failed to provide any one party with overall assembly majority. President Djohar appointed Halidi Abderamane Ibrahim prime minister.

## Congo
Republic of
(*République du Congo*)

***area*** 342,000 sq km/132,012 sq mi
***capital*** Brazzaville
***towns*** Pointe-Noire, N'Kayi, Loubomo
***physical*** narrow coastal plain rises to central plateau, then falls into northern basin; Zaïre (Congo) River on the border with Zaire; half the country is rainforest
***environment*** an estimated 93% of the rural population is without access to safe drinking water
***head of state and government*** Pascal Lissouba from 1992
***political system*** emergent democracy
***exports*** timber, petroleum, cocoa, sugar
***currency*** CFA franc
***population*** (1990 est) 2,305,000 (chiefly Bantu); growth rate 2.6% p.a.
***life expectancy*** men 45, women 48
***languages*** French (official); many African languages
***religions*** animist 50%, Christian 48%, Muslim 2%
***literacy*** men 79%, women 55% (1985 est)
***GDP*** $2.1 bn (1983); $500 per head
***chronology***
*1910* Became part of French Equatorial Africa.
*1960* Achieved independence, with Abbé Youlou as first president.

*1963* Youlou forced to resign. New constitution approved, with Alphonse Massamba-Débat as president.
*1964* The Congo became a one-party state.
*1968* Military coup, led by Capt Marien Ngouabi, ousted Massamba-Débat.
*1970* A Marxist state, the People's Republic of the Congo, was announced, with the Congolese Labour Party (PCT) as the only legal party.
*1977* Ngouabi assassinated. Col Yhombi-Opango became president.
*1979* Yhombi-Opango handed over the presidency to the PCT, who chose Col Denis Sassou-Nguessou as his successor.
*1984* Sassou-Nguessou elected for another five-year term.
*1990* The PCT abandoned Marxist-Leninism and promised multiparty politics.
*1991* 1979 constitution suspended. Country renamed the Republic of Congo.
*1992* New constitution approved and multiparty elections held, giving Pan-African Union for Social Democracy (UPADS) the most assembly seats.

**Costa Rica**
Republic of
(*República de Costa Rica*)

*area* 51,100 sq km/19,735 sq mi
*capital* San José
*towns* Limón, Puntarenas
*physical* high central plateau and tropical coasts; Costa Rica was once entirely forested, containing an estimated 5% of the Earth's flora and fauna. By 1983 only 17% of the forest remained; half of the arable land had been cleared for cattle ranching, which led to landlessness, unemployment (except for 2,000 politically powerful families), and soil erosion; the massive environmental destruction also caused incalculable loss to the gene pool
*environment* one of the leading centres of conservation in Latin America, with more than 10% of the country protected by national parks, and tree replanting proceeding at a rate of 150 sq km/60 sq mi per year
*head of state and government* Rafael Calderón from 1990
*political system* liberal democracy
*exports* coffee, bananas, cocoa, sugar, beef
*currency* colón
*population* (1990 est) 3,032,000 (including 1,200 Guaymi Indians); growth rate 2.6% p.a.

*life expectancy* men 71, women 76
*language* Spanish (official)
*religion* Roman Catholic 95%
*literacy* men 94%, women 93% (1985 est)
*GDP* $4.3 bn (1986); $1,550 per head
*chronology*
*1821* Independence achieved from Spain.
*1949* New constitution adopted. National army abolished. José Figueres, cofounder of the National Liberation Party (PLN) elected president; he embarked on ambitious socialist programme.
*1958–73* Mainly conservative administrations.
*1974* PLN regained the presidency and returned to socialist policies.
*1978* Rodrigo Carazo, conservative, elected president. Sharp deterioration in the state of the economy.
*1982* Luis Alberto Monge (PLN) elected president. Harsh austerity programme introduced to rebuild the economy. Pressure from the USA to abandon neutral stance and condemn Sandinista regime in Nicaragua.
*1983* Policy of neutrality reaffirmed.
*1985* Following border clashes with Sandinista forces, a US-trained antiguerrilla guard formed.
*1986* Oscar Arias won the presidency on a neutralist platform.
*1987* Arias won Nobel Prize for Peace for devising a Central American peace plan.
*1990* Rafael Calderón, Christian Socialist Union Party (PUSC), elected president.

# Croatia
Republic of

***area*** 56,538 sq km/21,824 sq mi
***capital*** Zagreb
***towns*** Rijeka (Fiume), Zadar, Sibenik, Split, Dubrovnik
***physical*** Adriatic coastline with large islands; very mountainous, with part of the Karst region and the Julian and Styrian Alps
***head of state*** Franjo Tudjman from 1990
***head of government*** Hrvoje Sarinic from 1992
***political system*** emergent democracy
***products*** cereals, potatoes, tobacco, fruit, livestock, metal goods, textiles
***currency*** Croatian dinar
***population*** (1990) 4,760,000 including 75% Croats, 12% Serbs, and 1% Slovenes
***language*** Croatian variant of Serbo-Croatian
***religions*** Roman Catholic (Croats); Orthodox Christian (Serbs)
***GNP*** $7.9 bn (1990); $1,660 per head
***chronology***
***1918*** Became part of the kingdom that united the Serbs, Croats,

and Slovenes.
*1929* The kingdom of Croatia, Serbia, and Slovenia became Yugoslavia. Croatia continued its campaign for autonomy.
*1941* Became a Nazi puppet state following German invasion.
*1945* Became constituent republic of Yugoslavia.
*1970s* Separatist demands resurfaced. Crackdown against anti-Serb separatist agitators.
*1989* Formation of opposition parties permitted.
*1990* April–May: Communists defeated by Tudjman-led Croatian Democratic Union (HDZ) in first free election since 1938.
Sept: 'sovereignty' declared. Dec: new constitution adopted.
*1991* Feb: assembly called for Croatia's secession. March: Serb-dominated Krajina announced secession from Croatia. June: Croatia declared independence; military conflict with Serbia; internal civil war ensued. July onwards: civil war intensified. Oct: Croatia formally seceded from Yugoslavia.
*1992* Jan: United Nations peace accord reached in Sarajevo; Croatia's independence recognized by the European Community. March–April: UN peacekeeping forces drafted into Croatia. April: independence recognized by USA. May: became a member of the United Nations. Aug: Tudjman directly elected president; HDZ won assembly elections. Sept: Tudjman requested withdrawal of UN forces on expiry of mandate 1993.
*1993* Jan: Croatian forces launched offensive to retake parts of Serb-held Krajina, violating the 1992 UN peace accord.

## Cuba
Republic of
(*República de Cuba*)

*area* 110,860 sq km/42,820 sq mi
*capital* Havana
*towns* Santiago de Cuba, Camagüey
*physical* comprises Cuba, the largest and westernmost of the West Indies, and smaller islands including Isle of Youth; low hills; Sierra Maestra mountains in SE
*head of state and government* Fidel Castro Ruz from 1959
*political system* communist republic
*exports* sugar, tobacco, coffee, nickel, fish
*currency* Cuban peso
*population* (1990 est) 10,582,000; 37% are white of Spanish descent, 51% mulatto, and 11% are of African origin; growth rate 0.6% p.a.
*life expectancy* men 72, women 75
*language* Spanish
*religions* Roman Catholic 85%; also Episcopalians and Methodists
*literacy* men 96%, women 95% (1988)
*disposable national income* $15.8 bn (1983); $1,590 per head
*chronology*
*1901* Cuba achieved independence; Tomás Estrada Palma became first

president of the Republic of Cuba.
*1933* Fulgencia Batista seized power.
*1944* Batista retired.
*1952* Batista seized power again to begin an oppressive regime.
*1953* Fidel Castro led an unsuccessful coup against Batista.
*1956* Second unsuccessful coup by Castro.
*1959* Batista overthrown by Castro. Constitution of 1940 replaced by a 'Fundamental Law', making Castro prime minister, his brother Raúl Castro his deputy, and 'Che' Guevara his number three.
*1960* All US businesses in Cuba appropriated without compensation; USA broke off diplomatic relations.
*1961* USA sponsored an unsuccessful invasion at the Bay of Pigs. Castro announced that Cuba had become a communist state, with a Marxist-Leninist programme of economic development.
*1962* Cuba expelled from the Organization of American States. Soviet nuclear missiles installed but subsequently removed from Cuba at US insistence.
*1965* Cuba's sole political party renamed Cuban Communist Party (PCC). With Soviet help, Cuba began to make considerable economic and social progress.
*1972* Cuba became a full member of the Moscow-based Council for Mutual Economic Assistance.
*1976* New socialist constitution approved; Castro elected president.
*1976–81* Castro became involved in extensive international commitments, sending troops as Soviet surrogates, particularly to Africa.
*1982* Cuba joined other Latin American countries in giving moral support to Argentina in its dispute with Britain over the Falklands.
*1984* Castro tried to improve US-Cuban relations by discussing exchange of US prisoners in Cuba for Cuban 'undesirables' in the USA.
*1988* Peace accord with South Africa signed, agreeing to withdrawal of Cuban troops from Angola.
*1989* Reduction in Cuba's overseas military activities.
*1991* Soviet troops withdrawn.
*1992* Castro affirmed continuing support of communism.

# Cyprus

**Cyprus**
Greek *Republic of Cyprus* (*Kypriakí Dimokratía*) in the south, and *Turkish Republic of Northern Cyprus* (*Kibris Cumhuriyeti*) in the north

*area* 9,251 sq km/3,571 sq mi, 37% in Turkish hands
*capital* Nicosia (divided between Greeks and Turks)
*towns* Limassol, Larnaca, Paphos (Greek); Morphou, Kyrenia, Famagusta (Turkish)
*physical* central plain between two E–W mountain ranges
*heads of state and government* Glafkos Clerides (Greek) from 1993, Rauf Denktaş (Turkish) from 1976
*political system* democratic divided republic
*exports* citrus, grapes, raisins, Cyprus sherry, potatoes, clothing, footwear
*currency* Cyprus pound and Turkish lira
*population* (1990 est) 708,000 (Greek Cypriot 78%, Turkish Cypriot 18%); growth rate 1.2% p.a.
*life expectancy* men 72, women 76
*languages* Greek and Turkish (official), English
*religions* Greek Orthodox 78%, Sunni Muslim 18%
*literacy* 99% (1984)
*GDP* $3.7 bn (1987); $5,497 per head

## Cyprus

*chronology*

**1878** Came under British administration.

**1955** Guerrilla campaign began against the British for enosis (union with Greece), led by Archbishop Makarios and General Grivas.

**1956** Makarios and enosis leaders deported.

**1959** Compromise agreed and Makarios returned to be elected president of an independent Greek–Turkish Cyprus.

**1960** Independence achieved from Britain, with Britain retaining its military bases.

**1963** Turks set up their own government in northern Cyprus. Fighting broke out between the two communities.

**1964** United Nations peacekeeping force installed.

**1971** Grivas returned to start a guerrilla war against the Makarios government.

**1974** Grivas died. Military coup deposed Makarios, who fled to Britain. Nicos Sampson appointed president. Turkish army sent to northern Cyprus to confirm Turkish Cypriots' control; military regime in southern Cyprus collapsed; Makarios returned. Northern Cyprus declared itself the Turkish Federated State of Cyprus (TFSC), with Rauf Denktaş as president.

**1977** Makarios died; succeeded by Spyros Kyprianou.

**1983** An independent Turkish Republic of Northern Cyprus proclaimed but recognized only by Turkey.

**1984** UN peace proposals rejected.

**1985** Summit meeting between Kyprianou and Denktaş failed to reach agreement.

**1988** Georgios Vassiliou elected president. Talks with Denktaş began, under UN auspices.

**1989** Vassiliou and Denktaş agreed to draft an agreement for the future reunification of the island, but peace talks were abandoned Sept.

**1991** Turkish offer of peace talks rejected by Cyprus and Greece.

**1992** UN-sponsored peace talks collapsed.

**1993** Glafkos Clerides replaced Vassiliou as Greek president.

# Czech Republic
(*Ceská Republika*)

*area* 78,864 sq km/30,461 sq mi
*capital* Prague
*towns* Brno, Ostrava, Olomouc, Liberec, Plzen, Ustí nad Labem
*physical* mountainous; rivers: Morava, Labe (Elbe), Vltava (Moldau)
*environment* considered in 1991 to be the most polluted country in E Europe. Pollution is worst in N Bohemia, which produces 70% of the country's coal and 45% of its coal-generated electricity. Up to 20 times the permissible level of sulphur dioxide is released over Prague, where 75% of the drinking water fails to meet the country's health standards
*head of state* Václav Havel from 1993
*head of government* Václav Klaus from 1993
*political system* emergent democracy
*exports* machinery, vehicles, coal, iron and steel, chemicals, glass, ceramics, clothing
*currency* new currency based on koruna
*population* (1991) 10,298,700 (with German and other minorities);

growth rate 0.4% p.a.
*life expectancy* men 68, women 75
*languages* Czech (official)
*religions* Roman Catholic 75%, Protestant, Hussite, Orthodox
*literacy* 100%
*GDP* $26,600 million (1990); $2,562 per head
*chronology*
*1526–1918* Under Habsburg domination.
*1918* Independence achieved from Austro-Hungarian Empire; Czechs joined Slovaks in forming Czechoslovakia as independent nation.
*1948* Communists assumed power in Czechoslovakia.
*1968* Czech Socialist Republic created under new federal constitution.
*1989* Nov: pro-democracy demonstrations in Prague; new political parties formed, including Czech-based Civic Forum under Václav Havel; Communist Party stripped of powers; political parties legalized. Dec: new 'grand coalition' government formed, including former dissidents; Havel appointed state president. Amnesty granted to 22,000 prisoners; calls for USSR to withdraw troops.
*1990* July: Havel re-elected president in multiparty elections.
*1991* Civic Forum split into Civil Democratic Party (CDP) and Civic Movement (CM); evidence of increasing Czech and Slovak separatism.
*1992* June: Václav Klaus, leader of the Czech-based CDP, became prime minister; Havel resigned following Slovak gains in assembly elections. Aug: creation of separate Czech and Slovak states agreed.
*1993* Jan: Czech Republic became sovereign state, with Klaus as prime minister. Havel elected president of the new republic. Admitted into United Nations, Conference on Security and Cooperation in Europe, and Council of Europe.

## Denmark
Kingdom of
(*Kongeriget Danmark*)

*area* 43,075 sq km/16,627 sq mi
*capital* Copenhagen
*towns* Aarhus, Odense, Aalborg, Esbjerg
*physical* comprises the Jutland peninsula and about 500 islands (100 inhabited) including Bornholm in the Baltic Sea; the land is flat and cultivated; sand dunes and lagoons on the W coast and long inlets (fjords) on the E; the main island is Sjælland (Zealand), where most of Copenhagen is located (the rest is on the island of Amager)
*territories* the dependencies of Faeroe Islands and Greenland
*head of state* Queen Margrethe II from 1972
*head of government* Poul Nyrup Rasmussen from 1993
*political system* liberal democracy
*exports* bacon, dairy produce, eggs, fish, mink pelts, car and aircraft parts, electrical equipment, textiles, chemicals
*currency* kroner
*population* (1990 est) 5,134,000; growth rate 0% p.a.
*life expectancy* men 72, women 78
*languages* Danish (official); there is a German-speaking minority
*religion* Lutheran 97%
*literacy* 99% (1983)
*GDP* $85.5 bn (1987); $16,673 per head

## Denmark

### recent chronology

**1940–45** Occupied by Germany.
**1945** Iceland's independence recognized.
**1947** Frederik IX succeeded Christian X.
**1948** Home rule granted for Faeroe Islands.
**1949** Became a founding member of NATO.
**1960** Joined European Free Trade Association (EFTA).
**1972** Margrethe II became Denmark's first queen in nearly 600 years.
**1973** Left EFTA and joined European Economic Community (EEC).
**1979** Home rule granted for Greenland.
**1985** Strong non-nuclear movement in evidence.
**1990** General election; another coalition government formed.
**1992** Rejection of Maastricht Treaty in national referendum; government requested modifications (codicils) to treaty prior to second national referendum, planned for 1993.
**1993** Poul Schlüter resigned; replaced by Poul Nyrup Rasmussen at head of Social Democrat-led coalition government.

# Djibouti

Republic of
(*Jumhouriyya Djibouti*)

*area* 23,200 sq km/8,955 sq mi
*capital* (and chief port) Djibouti
*towns* Tadjoura, Obock, Dikhil
*physical* mountains divide an inland plateau from a coastal plain; hot and arid
*head of state and government* Hassan Gouled Aptidon from 1977
*political system* authoritarian nationalism
*exports* acts mainly as a transit port for Ethiopia
*currency* Djibouti franc
*population* (1990 est) 337,000 (Issa 47%, Afar 37%, European 8%, Arab 6%); growth rate 3.4% p.a.
*life expectancy* 50
*languages* French (official), Somali, Afar, Arabic
*religion* Sunni Muslim
*literacy* 20% (1988)
*GDP* $378 million (1987); $1,016 per head
*chronology*
*1884* Annexed by France as part of French Somaliland.
*1967* French Somaliland became the French Territory of the Afars and the Issas.
*1977* Independence achieved from France; Hassan Gouled was

elected president.
*1979* All political parties combined to form the People's Progress Assembly (RPP).
*1981* New constitution made RPP the only legal party. Gouled re-elected. Treaties of friendship signed with Ethiopia, Somalia, Kenya, and Sudan.
*1984* Policy of neutrality reaffirmed.
*1987* Gouled re-elected for a final term.
*1991* Amnesty International accused secret police of brutality.
*1992* Djibouti elected member of UN Security Council 1993–95.

# Dominica
Commonwealth of

*area* 751 sq km/290 sq mi
*capital* Roseau, with a deepwater port
*towns* Portsmouth, Marigot
*physical* second largest of the Windward Islands, mountainous central ridge with tropical rainforest
*head of state* Clarence Seignoret from 1983
*head of government* Eugenia Charles from 1980
*political system* liberal democracy
*exports* bananas, coconuts, citrus, lime, bay oil
*currency* Eastern Caribbean dollar, pound sterling, French franc
*population* (1990 est) 94,200 (mainly black African in origin, but with a small Carib reserve of some 500); growth rate 1.3% p.a.
*life expectancy* men 57, women 59
*language* English (official), but the Dominican patois reflects earlier periods of French rule
*religion* Roman Catholic 80%
*literacy* 80%
*GDP* $91 million (1985); $1,090 per head
*chronology*
*1763* Became British possession.

*1978* Independence achieved from Britain. Patrick John, leader of Dominica Labour Party (DLP), elected prime minister.
*1980* Dominica Freedom Party (DFP), led by Eugenia Charles, won convincing victory in general election.
*1981* Patrick John implicated in plot to overthrow government.
*1982* John tried and acquitted.
*1985* John retried and found guilty. Regrouping of left-of-centre parties resulted in new Labour Party of Dominica (LPD). DFP, led by Eugenia Charles, re-elected.
*1990* Charles elected to a third term.
*1991* Integration into Windward Islands confederation proposed.

# Dominican Republic
(*República Dominicana*)

*area* 48,442 sq km/18,700 sq mi
*capital* Santo Domingo
*towns* Santiago de los Caballeros, San Pedro de Macoris
*physical* comprises eastern two-thirds of island of Hispaniola; central mountain range with fertile valleys
*head of state and government* Joaquín Ricardo Balaguer from 1986
*political system* democratic republic
*exports* sugar, gold, silver, tobacco, coffee, nickel
*currency* peso
*population* (1989 est) 7,307,000; growth rate 2.3% p.a.
*life expectancy* men 61, women 65
*language* Spanish (official)
*religion* Roman Catholic 95%
*literacy* men 78%, women 77% (1985 est)
*GDP* $4.9 bn (1987); $731 per head
*chronology*
*1844* Dominican Republic established.
*1930* Military coup established dictatorship of Rafael Trujillo.
*1961* Trujillo assassinated.
*1962* First democratic elections resulted in Juan Bosch, founder of the

Dominican Revolutionary Party (PRD), becoming president.
*1963* Bosch overthrown in military coup.
*1965* US Marines intervene to restore order and protect foreign nationals.
*1966* New constitution adopted. Joaquín Balaguer, leader of Christian Social Reform Party (PRSC), became president.
*1978* PRD returned to power, with Silvestre Antonio Guzmán as president.
*1982* PRD re-elected, with Jorge Blanco as president.
*1985* Blanco forced by International Monetary Fund to adopt austerity measures to save the economy.
*1986* PRSC returned to power, with Balaguer re-elected president.

## Ecuador
Republic of
(*República del Ecuador*)

*area* 270,670 sq km/104,479 sq mi
*capital* Quito
*towns* Cuenca, Guayaquil
*physical* coastal plain rises sharply to Andes Mountains, which are divided into a series of cultivated valleys; low-lying rainforest in E
*environment* about 25,000 species became extinct 1965–90 as a result of environmental destruction
*head of state and government* Sixto Duran Ballen from 1992
*political system* emergent democracy
*exports* bananas, cocoa, coffee, sugar, rice, fruit, fish, petroleum
*currency* sucre
*population* (1989 est) 10,490,000; (mestizo 55%, Indian 25%, European 10%, black African 10%); growth rate 2.9% p.a.
*life expectancy* men 62, women 66
*languages* Spanish (official); Quechua, Jivaro, and other Indian languages

***religion*** Roman Catholic 95%
***literacy*** men 85%, women 80% (1985 est)
***GDP*** $10.6 bn (1987); $1,069 per head
***chronology***
***1830*** Independence achieved from Spain.
***1925–48*** Great political instability; no president completed his term of office.
***1948–55*** Liberals in power.
***1956*** First conservative president in 60 years.
***1960*** Liberals returned, with José Velasco as president.
***1961*** Velasco deposed and replaced by the vice president.
***1962*** Military junta installed.
***1968*** Velasco returned as president.
***1972*** A coup put the military back in power.
***1978*** New democratic constitution adopted.
***1979*** Liberals in power but opposed by right- and left-wing parties.
***1982*** Deteriorating economy provoked strikes, demonstrations, and a state of emergency.
***1983*** Austerity measures introduced.
***1984–85*** No party with a clear majority in the national congress; Febres Cordero narrowly won the presidency for the Conservatives.
***1988*** Rodrigo Borja Cevallos elected president for moderate left-wing coalition.
***1989*** Guerrilla left-wing group, *Alfaro Vive, Carajo* ('Alfaro lives, Dammit'), numbering about 1,000, laid down arms after nine years.
***1992*** United Republican Party (PUR) leader, Sixto Duran Ballen, elected president; Social Christian Party (PSC) became largest party in congress.

# Egypt
Arab Republic of
(*Jumhuriyat Misr al-Arabiya*)

*area* 1,001,450 sq km/386,990 sq mi
*capital* Cairo
*towns* Gîza, Alexandria, Port Said, Suez, Damietta
*physical* mostly desert; hills in E; fertile land along Nile valley and delta; cultivated and settled area is about 35,500 sq km/13,700 sq mi
*environment* the building of the Aswan Dam (opened 1970) on the Nile has caused widespread salinization and an increase in waterborne diseases in villages close to Lake Nasser. A dramatic fall in the annual load of silt deposited downstream has reduced the fertility of cropland and has led to coastal erosion and the consequent loss of sardine shoals
*head of state and government* Hosni Mubarak from 1981
*political system* democratic republic
*exports* cotton and textiles, petroleum, fruit and vegetables
*currency* Egyptian pound
*population* (1989 est) 54,779,000; growth rate 2.4% p.a.
*life expectancy* men 57, women 60
*languages* Arabic (official); ancient Egyptian survives to some extent in Coptic
*religions* Sunni Muslim 95%, Coptic Christian 5%
*literacy* men 59%, women 30% (1985 est)
*GDP* $34.5 bn (1987); $679 per head

## chronology

**1914** Egypt became a British protectorate.
**1936** Independence achieved from Britain. King Fuad succeeded by his son Farouk.
**1946** Withdrawal of British troops except from Suez Canal Zone.
**1952** Farouk overthrown by army in bloodless coup.
**1953** Egypt declared a republic, with General Neguib as president.
**1956** Neguib replaced by Col Gamal Nasser. Nasser announced nationalization of Suez Canal; Egypt attacked by Britain, France, and Israel. Cease-fire agreed because of US intervention.
**1958** Short-lived merger of Egypt and Syria as United Arab Republic (UAR). Subsequent attempts to federate Egypt, Syria, and Iraq failed.
**1967** Six-Day War with Israel ended in Egypt's defeat and Israeli occupation of Sinai and Gaza Strip.
**1970** Nasser died suddenly; succeeded by Anwar Sadat.
**1973** Attempt to regain territory lost to Israel led to fighting; cease-fire arranged by US secretary of state Henry Kissinger.
**1977** Sadat's visit to Israel to address the Israeli parliament was criticized by Egypt's Arab neighbours.
**1978–79** Camp David talks in the USA resulted in a treaty between Egypt and Israel. Egypt expelled from the Arab League.
**1981** Sadat assassinated, succeeded by Hosni Mubarak.
**1983** Improved relations between Egypt and the Arab world; only Libya and Syria maintained a trade boycott.
**1984** Mubarak's party victorious in the people's assembly elections.
**1987** Mubarak re-elected. Egypt readmitted to Arab League.
**1988** Full diplomatic relations with Algeria restored.
**1989** Improved relations with Libya; diplomatic relations with Syria restored. Mubarak proposed a peace plan.
**1991** Participation in Gulf War on US-led side. Major force in convening Middle East peace conference in Spain.
**1992** Outbreaks of violence between Muslims and Christians. Earthquake devastated Cairo.

# El Salvador
Republic of
(*República de El Salvador*)

*area* 21,393 sq km/8,258 sq mi
*capital* San Salvador
*towns* Santa Ana, San Miguel
*physical* narrow coastal plain, rising to mountains in N with central plateau
*head of state and government* Alfredo Cristiani from 1989
*political system* emergent democracy
*exports* coffee, cotton, sugar
*currency* colón
*population* (1989 est) 5,900,000 (mainly of mixed Spanish and Indian ancestry; 10% Indian); growth rate 2.9% p.a.
*life expectancy* men 63, women 66
*languages* Spanish, Nahuatl
*religion* Roman Catholic 97%
*literacy* men 75%, women 69% (1985 est)
*GDP* $4.7 bn (1987); $790 per head
*chronology*
*1821* Independence achieved from Spain.
*1931* Peasant unrest followed by a military coup.

*1961* Following a coup, the conservative National Conciliation Party (PCN) established and in power.
*1969* 'Soccer' war with Honduras.
*1972* Allegations of human-rights violations; growth of left-wing guerrilla activities. General Carlos Romero elected president.
*1979* A coup replaced Romero with a military–civilian junta.
*1980* Archbishop Oscar Romero assassinated; country on verge of civil war. José Duarte became first civilian president since 1931.
*1981* Mexico and France recognized the guerrillas as a legitimate political force, but the USA actively assisted the government in its battle against them.
*1982* Assembly elections boycotted by left-wing parties and held amid considerable violence.
*1986* Duarte sought a negotiated settlement with the guerrillas.
*1988* Duarte resigned.
*1989* Alfredo Cristiani, National Republican Alliance (ARENA), became president in rigged elections; rebel attacks intensified.
*1991* UN-sponsored peace accord signed by representatives of the government and the socialist guerrilla group, the Farabundo Marti Liberation Front (FMLN).
*1992* Peace accord validated; FMLN became political party.

# Equatorial Guinea
Republic of
(*República de Guinea Ecuatorial*)

*area* 28,051 sq km/10,828 sq mi
*capital* Malabo (Bioko)
*towns* Bata, Mbini (Río Muni)
*physical* comprises mainland Río Muni, plus the small islands of Corisco, Elobey Grande and Elobey Chico, and Bioko (formerly Fernando Po) together with Annobón (formerly Pagalu)
*head of state and government* Teodoro Obiang Nguema Mbasogo from 1979
*political system* one-party military republic
*exports* cocoa, coffee, timber
*currency* ekuele; CFA franc
*population* (1988 est) 336,000 (plus 110,000 estimated to live in exile abroad); growth rate 2.2% p.a.
*life expectancy* men 44, women 48
*languages* Spanish (official); pidgin English is widely spoken, and on Annobón (whose people were formerly slaves of the Portuguese) a Portuguese dialect; Fang and other African dialects spoken on Río Muni
*religions* nominally Christian, mainly Catholic, but in 1978 Roman Catholicism was banned
*literacy* 55% (1984)

***GDP*** $90 million (1987); $220 per head
*chronology*
*1778* Fernando Po (Bioko Island) ceded to Spain.
*1885* Mainland territory came under Spanish rule; colony known as Spanish Guinea.
*1968* Independence achieved from Spain. Francisco Macias Nguema became first president, soon assuming dictatorial powers.
*1979* Macias overthrown and replaced by his nephew, Teodoro Obiang Nguema Mbasogo, who established a military regime. Macias tried and executed.
*1982* Obiang elected president unopposed for another seven years. New constitution adopted.
*1989* Obiang re-elected president.
*1992* New constitution adopted; elections held, but president continued to nominate candidates for top government posts.

# Estonia
Republic of

*area* 45,000 sq km/17,000 sq mi
*capital* Tallinn
*towns* Tartu, Narva, Kohtla-Järve, Pärnu
*physical* lakes and marshes in a partly forested plain; 774 km/481 mi of coastline; mild climate
*head of state* Lennart Meri from 1992
*head of government* Tiit Vahl from 1992
*political system* emergent democracy
*products* oil and gas (from shale), wood products, flax, dairy and pig products
*currency* kroon
*population* (1989) 1,573,000 (Estonian 62%, Russian 30%, Ukrainian 3%, Byelorussian 2%)
*language* Estonian, allied to Finnish
*religion* traditionally Lutheran
*chronology*
*1918* Estonia declared its independence. March: Soviet forces, who had tried to regain control from occupying German forces during World War I, were overthrown by German troops. Nov: Soviet troops took control after German withdrawal.

*1919* Soviet rule overthrown with help of British navy; Estonia declared a democratic republic.
*1934* Fascist coup replaced government.
*1940* Estonia incorporated into USSR.
*1941–44* German occupation during World War II.
*1944* USSR regained control.
*1980* Beginnings of nationalist dissent.
*1988* Adopted own constitution, with power of veto on all centralized Soviet legislation. Popular Front (Rahvarinne) established to campaign for democracy. Estonia's supreme soviet (state assembly) voted to declare the republic 'sovereign' and autonomous in all matters except military and foreign affairs; rejected by USSR as unconstitutional.
*1989* Estonian replaced Russian as main language.
*1990* Feb: Communist Party monopoly of power abolished; multiparty system established. March: pro-independence candidates secured majority after republic elections; coalition government formed with Popular Front leader Edgar Savisaar as prime minister; Arnold Rüütel became president. May: prewar constitution partially restored.
*1991* March: independence plebiscite overwhelmingly approved. Aug: full independence declared after abortive anti-Gorbachev coup; Communist Party outlawed. Sept: independence recognized by Soviet government and Western nations; admitted into United Nations and CSCE (Conference on Security and Cooperation in Europe).
*1992* Jan: Savisaar resigned owing to his government's inability to alleviate food and energy shortages; new government formed by Tiit Vahl. June: New constitution approved. Sept: presidential election inconclusive; right-wing Fatherland Group did well in general election. Oct: Meri chosen by parliament to replace Rüütel.

# Ethiopia

**Ethiopia** People's Democratic Republic of (*Hebretesebawit Ityopia*, formerly also known as *Abyssinia*)

*area* 1,221,900 sq km/471,653 sq mi
*capital* Addis Ababa
*towns* Asmara (capital of Eritrea), Dire Dawa, Massawa, Assab
*physical* a high plateau with central mountain range divided by Rift Valley; plains in E; source of Blue Nile River
*environment* more than 90% of the forests of the Ethiopian highlands have been destroyed since 1900
*head of state and government* Meles Zenawi from 1991
*political system* transition to democratic socialist republic
*exports* coffee, pulses, oilseeds, hides, skins
*currency* birr
*population* (1989 est) 47,709,000 (Oromo 40%, Amhara 25%, Tigré 12%, Sidamo 9%); growth rate 2.5% p.a.
*life expectancy* 38
*languages* Amharic (official), Tigrinya, Orominga, Arabic
*religions* Sunni Muslim 45%, Christian (Ethiopian Orthodox Church, which has had its own patriarch since 1976) 40%
*literacy* 35% (1988)
*GDP* $4.8 bn (1987); $104 per head
*chronology*
*1889* Abyssinia reunited by Menelik II.

*1930* Haile Selassie became emperor.

*1962* Eritrea annexed by Haile Selassie; resistance movement began.

*1974* Haile Selassie deposed and replaced by a military government led by General Teferi Benti. Ethiopia declared a socialist state.

*1977* Teferi Benti killed and replaced by Col Mengistu Haile Mariam.

*1977–79* 'Red Terror' period in which Mengistu's regime killed thousands of people.

*1981–85* Ethiopia spent at least $2 billion on arms.

*1985* Worst famine in more than a decade; Western aid sent and forcible internal resettlement programmes undertaken.

*1987* New constitution adopted, Mengistu Mariam elected president. New famine; food aid hindered by guerrillas.

*1988* Mengistu agreed to adjust his economic policies in order to secure International Monetary Fund assistance. Influx of refugees from Sudan.

*1989* Coup attempt against Mengistu foiled. Peace talks with Eritrean rebels mediated by former US president Carter reported some progress.

*1990* Rebels captured port of Massawa. Mengistu announced new reforms.

*1991* Mengistu overthrown; transitional government set up by Ethiopian People's Revolutionary Democratic Front (EPRDF). Eritrean People's Liberation Front (EPLF) secured Eritrea; Eritrea's right to secede recognized; Meles Zenawi elected Ethiopia's new head of state and government. Isaias Afwerki became secretary general of provisional government in Eritrea.

*1993* April: overwhelming majority voted in favour of Eritrean independence in referendum.

# Fiji
Republic of

*area* 18,333 sq km/7,078 sq mi
*capital* Suva
*towns* Lautoka, Levuka
*physical* comprises 844 Melanesian and Polynesian islands and islets (about 110 inhabited), the largest being Viti Levu (10,429 sq km/4,028 sq mi) and Vanua Levu (5,550 sq km/2,146 sq mi); mountainous, volcanic, with tropical rainforest and grasslands
*head of state* Ratu Sir Penaia Ganilau from 1987
*head of government* Col Sitiveni Rabuka from 1992
*political system* democratic republic
*exports* sugar, coconut oil, ginger, timber, canned fish, gold; tourism is important
*currency* Fiji dollar
*population* (1989 est) 758,000 (46% Fijian, holding 80% of the land communally, and 49% Indian, introduced in the 19th century to work the sugar crop); growth rate 2.1% p.a.
*life expectancy* men 67, women 71
*languages* English (official), Fijian, Hindi
*religions* Hindu 50%, Methodist 44%
*literacy* men 88%, women 77% (1980 est)
*GDP* $1.2 bn (1987); $1,604 per head

*chronology*
*1874* Fiji became a British crown colony.
*1970* Independence achieved from Britain; Ratu Sir Kamisese Mara elected as first prime minister.
*1987* April: general election brought to power an Indian-dominated coalition led by Dr Timoci Bavadra. May: military coup by Col Sitiveni Rabuka removed new government at gunpoint; Governor General Ratu Sir Penaia Ganilau regained control within weeks. Sept: second military coup by Rabuka proclaimed Fiji a republic and suspended the constitution. Oct: Fiji ceased to be a member of the Commonwealth. Dec: civilian government restored with Rabuka retaining control of security as minister for home affairs.
*1990* New constitution, favouring indigenous Fijians, introduced.
*1992* General election produced coalition government; Col Rabuka named as president.

# Finland
Republic of
(*Suomen Tasavalta*)

*area* 338,145 sq km/ 130,608 sq mi
*capital* Helsinki
*towns* Tampere, Rovaniemi, Lahti; ports Turku, Oulu
*physical* most of the country is forest, with low hills and about 60,000 lakes; one-third is within the Arctic Circle; archipelago in S; includes Åland Islands
*head of state* Mauno Koivisto from 1982
*head of government* Esko Aho from 1991
*political system* democratic republic
*exports* metal, chemical, and engineering products (icebreakers and oil rigs), paper, sawn wood, clothing, fine ceramics, glass, furniture
*currency* markka
*population* (1989 est) 4,990,000; growth rate 0.5% p.a.
*life expectancy* men 70, women 78
*languages* Finnish 93%, Swedish 6% (both official), small Saami- and Russian-speaking minorities
*religions* Lutheran 97%, Eastern Orthodox 1.2%
*literacy* 99%
*GDP* $77.9 bn (1987); $15,795 per head
*chronology*
*1809* Finland annexed by Russia.

*1917* Independence declared from Russia.
*1920* Soviet regime acknowledged independence.
*1939* Defeated by USSR in Winter War.
*1941* Allowed Germany to station troops in Finland to attack USSR; USSR bombed Finland.
*1944* Concluded separate armistice with USSR.
*1948* Finno-Soviet Pact of Friendship, Cooperation, and Mutual Assistance signed.
*1955* Finland joined the United Nations and the Nordic Council.
*1956* Urho Kekkonen elected president; re-elected 1962, 1968, 1978.
*1973* Trade treaty with European Economic Community signed.
*1977* Trade agreement with USSR signed.
*1982* Mauno Koivisto elected president; re-elected 1988.
*1989* Finland joined Council of Europe.
*1991* Big swing to the centre in general election. New coalition government formed.
*1992* Formal application for European Community membership.

## France

French Republic
(*République Française*)

*area* (including island of Corsica) 543,965 sq km/ 209,970 sq mi
*capital* Paris
*towns* Lyons, Bordeaux, Toulouse, Strasbourg, Marseille, Nice
*physical* rivers Seine, Loire, Garonne, Rhône, Rhine; mountain ranges Alps, Massif Central, Pyrenees, Jura, Vosges, Cévennes
*territories* Guadeloupe, French Guiana, Martinique, Réunion, St Pierre and Miquelon, Southern and Antarctic Territories, New Caledonia, French Polynesia, Wallis and Futuna
*head of state* François Mitterrand from 1981
*head of government* Edouard Balladur from 1993
*political system* liberal democracy
*exports* fruit (especially apples), wine, cheese, wheat, cars, aircraft, iron and steel, petroleum products, chemicals, jewellery, silk, lace; tourism is very important
*currency* franc
*population* (1990 est) 56,184,000 (including 4,500,000 immigrants, chiefly from Portugal, Algeria, Morocco, and Tunisia); growth rate 0.3% p.a.

*life expectancy* men 71, women 79
*language* French; regional languages include Basque, Breton, Catalan, and the Provençal dialect
*religions* Roman Catholic 90%, Protestant 2%, Muslim 1%
*literacy* 99% (1984)
*GNP* $568 bn (1983); $7,179 per head
*recent chronology*
***1944–46*** Provisional government headed by General Charles de Gaulle; start of Fourth Republic.
***1954*** Indochina achieved independence.
***1956*** Morocco and Tunisia achieved independence.
***1957*** Entry into European Economic Community.
***1958*** Recall of de Gaulle after Algerian crisis; start of Fifth Republic.
***1959*** De Gaulle became president.
***1962*** Algeria achieved independence.
***1966*** France withdrew from military wing of NATO.
***1968*** 'May events' crisis.
***1969*** De Gaulle resigned after referendum defeat; Georges Pompidou became president.
***1974*** Giscard d'Estaing elected president.
***1981*** Mitterrand elected Fifth Republic's first socialist president.
***1986*** 'Cohabitation' experiment, with the conservative Jacques Chirac as prime minister.
***1988*** Mitterrand re-elected. Moderate socialist Michel Rocard became prime minister. Matignon Accord on future of New Caledonia approved by referendum.
***1989*** Greens gained 11% of vote in elections to European Parliament.
***1991*** French forces were part of the US-led coalition in the Gulf War. Edith Cresson became France's first woman prime minister. Mitterrand's popularity rating fell rapidly.
***1992*** March: Socialist Party humiliated in regional and local elections; Greens and National Front polled strongly. April: Cresson replaced by Pierre Bérégovoy. Sept: referendum narrowly endorsed the Maastricht Treaty.
***1993*** March: Socialist Party suffered heavy defeat in National Assembly elections. Edouard Balladur appointed prime minister.

## Gabon
Gabonese Republic
(*République Gabonaise*)

*area* 267,667 sq km/103,319 sq mi
*capital* Libreville
*towns* Port-Gentil, Owendo, Masuku (Franceville)
*physical* virtually the whole country is tropical rainforest; narrow coastal plain rising to hilly interior with savanna in E and S; Ogooué River flows N–W
*head of state and government* Omar Bongo from 1967
*political system* emergent democracy
*exports* petroleum, manganese, uranium, timber
*currency* CFA franc
*population* (1988) 1,226,000 including 40 Bantu groups; growth rate 1.6% p.a.
*life expectancy* men 47, women 51
*languages* French (official), Bantu
*religions* Christian 96% (Roman Catholic 65%), small Muslim minority 1%, animist 3%
*literacy* men 70%, women 53% (1985 est)
*GDP* $3.5 bn (1987); $3,308 per head
*chronology*
*1889* Gabon became part of the French Congo.
*1960* Independence from France; Léon M'ba became the first

president.
***1967*** Attempted coup by rival party foiled with French help. M'ba died; he was succeeded by his protégé Albert-Bernard Bongo.
***1968*** One-party state established.
***1973*** Bongo re-elected; converted to Islam, he changed his first name to Omar.
***1986*** Bongo re-elected.
***1989*** Coup attempt against Bongo defeated.
***1990*** Gabonese Democratic Party (PDG) won first multiparty elections since 1964 amidst allegations of ballot-rigging.

# Gambia
Republic of The

*area* 10,402 sq km/4,018 sq mi
*capital* Banjul
*towns* Serekunda, Bakau, Georgetown
*physical* banks of the river Gambia flanked by low hills
*head of state and government* Dawda K Jawara from 1970
*political system* liberal democracy
*exports* groundnuts, palm oil, fish
*currency* dalasi
*population* (1990 est) 820,000; growth rate 1.9% p.a.
*life expectancy* 42 (1988 est)
*languages* English (official), Mandinka, Fula and other native tongues
*religions* Muslim 90%, with animist and Christian minorities
*literacy* men 36%, women 15% (1985 est)
*GDP* $189 million (1987); $236 per head
*chronology*
*1843* The Gambia became a crown colony.
*1965* Independence achieved from Britain as a constitutional monarchy within the Commonwealth, with Dawda K Jawara as prime minister.
*1970* Declared itself a republic, with Jawara as president.
*1972* Jawara re-elected.

***1981*** Attempted coup foiled with the help of Senegal.
***1982*** Formed with Senegal the Confederation of Senegambia; Jawara re-elected.
***1987*** Jawara re-elected.
***1989*** Confederation of Senegambia dissolved.
***1990*** Gambian troops contributed to the stabilizing force in Liberia.

## Georgia
Republic of

*area* 69,700 sq km/26,911 sq mi
*capital* Tbilisi
*towns* Kutaisi, Rustavi, Batumi, Sukhumi
*physical* largely mountainous with a variety of landscape from the subtropical Black Sea shores to the ice and snow of the crest line of the Caucasus; chief rivers are Kura and Rioni
*interim head of state* Eduard Shevardnadze from 1992
*head of government* Tengiz Sigua from 1992
*political system* transitional
*products* tea, citrus and orchard fruits, tung oil, tobacco, vines, silk, hydroelectricity
*population* (1990) 5,500,000 (Georgian 70%, Armenian 8%, Russian 8%, Azeri 6%, Ossetian 3%, Abkhazian 2%)
*language* Georgian
*religion* Georgian Church, independent of the Russian Orthodox Church since 1917
*chronology*
*1918–21* Independent republic.
*1921* Uprising quelled by Red Army; Soviet republic established.
*1922–36* Linked with Armenia and Azerbaijan as the Transcaucasian Republic.

*1936* Became separate republic within USSR.
*1981–88* Increasing demands for autonomy, spearheaded from 1988 by the Georgian Popular Front.
*1989* March–April: Abkhazians demanded secession from Georgia, provoking inter-ethnic clashes. April: Georgian Communist Party (GCP) leadership purged. July: state of emergency imposed in Abkhazia; inter-ethnic clashes in South Ossetia. Nov: economic and political sovereignty declared.
*1990* March: GCP monopoly ended. Oct: nationalist coalition triumphed in supreme soviet elections. Nov: Zviad Gamsakhurdia became president. Dec: GCP seceded from Communist Party of USSR; calls for Georgian independence.
*1991* April: declared independence. May: Gamsakhurdia popularly elected president. Aug: GCP outlawed and all relations with USSR severed. Sept: anti-Gamsakhurdia demonstrations; state of emergency declared. Dec: Georgia failed to join new Commonwealth of Independent States (CIS).
*1992* Jan: Gamsakhurdia fled to Armenia; Sigua appointed prime minister; Georgia admitted into Conference on Security and Cooperation in Europe (CSCE). Eduard Shevardnadze appointed interim president. July: admitted into United Nations (UN).
Aug: fighting started between Georgian troops and Abkhazian separatists in Abkhazia in NW. Oct: Shevardnadze elected chair of new parliament. Clashes in South Ossetia and Abkhazia continued.
*1993* Inflation at 1,500%.

## Germany
Federal Republic of
(*Bundesrepublik Deutschland*)

*area* 357,041 sq km/137,853 sq mi
*capital* Berlin
*towns* Cologne, Munich, Essen, Frankfurt-am-Main, Dortmund, Hamburg
*physical* flat in N, mountainous in S with Alps; rivers Rhine, Weser, Elbe flow N, Danube flows SE, Oder, Neisse flow N along Polish frontier; many lakes, including Müritz
*environment* acid rain causing *Waldsterben* (tree death) affects more than half the country's forests; industrial E Germany has the highest sulphur-dioxide emissions in the world per head of population
*head of state* Richard von Weizsäcker from 1984
*head of government* Helmut Kohl from 1982
*political system* liberal democratic federal republic
*exports* machine tools (world's leading exporter), cars, commercial vehicles, electronics, industrial goods, textiles, chemicals, iron, steel, wine, lignite (world's largest producer), uranium, coal, fertilizers, plastics
*currency* Deutschmark

***population*** (1990) 78,420,000 (including nearly 5,000,000 'guest workers', *Gastarbeiter*, of whom 1,600,000 are Turks; the rest are Yugoslav, Italian, Greek, Spanish, and Portuguese); growth rate −0.7% p.a.
***life expectancy*** men 68, women 74
***languages*** German, Sorbian
***religions*** Protestant 42%, Roman Catholic 35%
***literacy*** 99% (1985)
***GNP*** $1,250 bn (1989); $16,200 per head

*recent chronology*
***1945*** Germany surrendered; country divided into four occupation zones (US, French, British, Soviet).
***1948*** Blockade of West Berlin.
***1949*** Establishment of Federal Republic under the 'Basic Law' Constitution with Konrad Adenauer as chancellor; establishment of the German Democratic Republic as an independent state.
***1953*** Uprising in East Berlin suppressed by Soviet troops.
***1954*** Grant of full sovereignty to both West Germany and East Germany.
***1957*** West Germany was a founder-member of the European Economic Community; recovery of the Saarland from France.
***1961*** Construction of Berlin Wall.
***1963*** Retirement of Chancellor Adenauer.
***1964*** Treaty of Friendship and Mutual Assistance signed between East Germany and USSR.
***1969*** Willy Brandt became chancellor of West Germany.
***1971*** Erich Honecker elected Socialist Unity Party (SED) leader in East Germany.
***1972*** Basic Treaty between West Germany and East Germany; treaty ratified 1973, normalizing relations between the two.
***1974*** Resignation of Brandt; Helmut Schmidt became chancellor.
***1975*** East German friendship treaty with USSR renewed for 25 years.
***1982*** Helmut Kohl became West German chancellor.
***1987*** Official visit of Honecker to the Federal Republic.
***1988*** Death of Franz-Josef Strauss, leader of the West German Bavarian Christian Social Union (CSU)

*1989* West Germany: rising support for far right in local and European elections, declining support for Kohl. East Germany: mass exodus to West Germany began. Honecker replaced by Egon Krenz. National borders opened in Nov, including Berlin Wall. Reformist Hans Modrow appointed prime minister. Krenz replaced.

*1990* March: East German multiparty elections won by a coalition led by the right-wing Christian Democratic Union (CDU). 3 Oct: official reunification of East and West Germany. 2 Dec: first all-German elections since 1932, resulting in a victory for Kohl.

*1991* Kohl's popularity declined after tax increase. The CDU lost its Bundesrat majority to the Social Democratic Party (SPD). Racism continued with violent attacks on foreigners.

*1992* Neo-Nazi riots against immigrants continued.

*1993* Jan: unemployment exceeded 7%; 1% decline in national output predicted for 1993. Honecker allowed to leave for exile in Chile. March: support for CDU slumped in state election in Hesse; Republicans captured 8% of vote.

# Ghana
Republic of

*area* 238,305 sq km/91,986 sq mi
*capital* Accra
*towns* Kumasi, ports Sekondi-Takoradi, Tema
*physical* mostly tropical lowland plains; bisected by river Volta
*environment* forested areas have shrunk from 8.2 million sq km/3.17 million sq mi at the beginning of the 20th century to 1.9 million sq km/730,000 sq mi by 1990
*head of state and government* Jerry Rawlings from 1981
*political system* military republic
*exports* cocoa, coffee, timber, gold, diamonds, manganese, bauxite
*currency* cedi
*population* (1990 est) 15,310,000; growth rate 3.2% p.a.
*life expectancy* men 50, women 54
*languages* English (official) and African languages
*religion* animist 38%, Muslim 30%, Christian 24%
*literacy* men 64%, women 43% (1985 est)
*GNP* $3.9 bn (1983); $420 per head
*chronology*
*1957* Independence achieved from Britain, within the Commonwealth, with Kwame Nkrumah as president.
*1960* Ghana became a republic.

*1964* Ghana became a one-party state.
*1966* Nkrumah deposed and replaced by General Joseph Ankrah.
*1969* Ankrah replaced by General Akwasi Afrifa, who initiated a return to civilian government.
*1970* Edward Akufo-Addo elected president.
*1972* Another coup placed Col Acheampong at the head of a military government.
*1978* Acheampong deposed in a bloodless coup led by Frederick Akuffo; another coup put Flight-Lt Jerry Rawlings in power.
*1979* Return to civilian rule under Hilla Limann.
*1981* Rawlings seized power again, citing the incompetence of previous governments. All political parties banned.
*1989* Coup attempt against Rawlings foiled.
*1992* New multiparty constitution approved. Partial lifting of ban on political parties. Nov: Rawlings won presidency in national elections.
*1993* Fourth republic of Ghana formally inaugurated in Rawlings's presence.

# Greece
Hellenic Republic
(*Elliniki Dimokratia*)

*area* 131,957 sq km/50,935 sq mi
*capital* Athens
*towns* Larisa, Piraeus, Thessaloníki, Patras, Iráklion
*physical* mountainous; a large number of islands, notably Crete, Corfu, and Rhodes
*environment* acid rain and other airborne pollutants are destroying the Classical buildings and ancient monuments of Athens
*head of state* Constantine Karamanlis from 1990
*head of government* Constantine Mitsotakis from 1990
*political system* democratic republic
*exports* tobacco, fruit, vegetables, olives, olive oil, textiles, aluminium, iron and steel
*currency* drachma
*population* (1990 est) 10,066,000; growth rate 0.3% p.a.
*life expectancy* men 72, women 76
*language* Greek
*religion* Greek Orthodox 97%
*literacy* men 96%, women 89% (1985)
*GDP* $40.9 bn (1987); $4,093 per head

*chronology*
*1829* Independence achieved from Turkish rule.
*1912–13* Balkan Wars; Greece gained much land.
*1941–44* German occupation of Greece.
*1946* Civil war between royalists and communists; communists defeated.
*1949* Monarchy re-established with Paul as king.
*1964* King Paul succeeded by his son Constantine.
*1967* Army coup removed the king; Col George Papadopoulos became prime minister. Martial law imposed, all political activity banned.
*1973* Republic proclaimed, with Papadopoulos as president.
*1974* Former premier Constantine Karamanlis recalled from exile to lead government. Martial law and ban on political parties lifted; restoration of the monarchy rejected by a referendum.
*1975* New constitution adopted, making Greece a democratic republic.
*1980* Karamanlis resigned as prime minister and was elected president.
*1981* Greece became full member of European Economic Community. Andreas Papandreou elected Greece's first socialist prime minister.
*1983* Five-year military and economic cooperation agreement signed with USA; ten-year economic cooperation agreement signed with USSR.
*1985* Papandreou re-elected.
*1988* Relations with Turkey improved. Major cabinet reshuffle after mounting criticism of Papandreou.
*1989* Papandreou defeated. Tzannis Tzannetakis became prime minister; his all-party government collapsed. Xenophon Zolotas formed new unity government. Papandreou charged with corruption.
*1990* New Democracy Party (ND) won half of parliamentary seats in general election but no outright majority; Constantine Mitsotakis became premier; formed new all-party government. Karamanlis re-elected president.
*1992* Papandreou acquitted. Greece opposed recognition of independence of the Yugoslav breakaway republic of Macedonia. Decisive parliamentary vote to ratify Maastricht Treaty.

# Grenada

*area* (including the Grenadines) 340 sq km/131 sq mi
*capital* St George's
*towns* Grenville, Hillsborough
*physical* southernmost of the Windward Islands; mountainous
*head of state* Elizabeth II from 1974, represented by governor general
*head of government* Nicholas Braithwaite from 1990
*political system* emergent democracy
*exports* cocoa, nutmeg, bananas, mace
*currency* Eastern Caribbean dollar
*population* (1990 est) 84,000, 84% of black African descent; growth rate –0.2% p.a.
*life expectancy* 69
*language* English (official); some French patois spoken
*religion* Roman Catholic 60%
*literacy* 85% (1985)
*GDP* $139 million (1987); $1,391 per head
*chronology*
*1974* Independence achieved from Britain; Eric Gairy elected prime minister.

*1979* Gairy removed in bloodless coup led by Maurice Bishop; constitution suspended and a People's Revolutionary Government established.
*1982* Relations with the USA and Britain deteriorated as ties with Cuba and the USSR strengthened.
*1983* After Bishop's attempt to improve relations with the USA, he was overthrown by left-wing opponents. A coup established the Revolutionary Military Council (RMC), and Bishop and three colleagues were executed. The USA invaded Grenada, accompanied by troops from other E Caribbean countries; RMC overthrown, 1974 constitution reinstated.
*1984* The newly formed New National Party (NNP) won 14 of the 15 seats in the house of representatives and its leader, Herbert Blaize, became prime minister.
*1989* Herbert Blaize lost leadership of NNP, remaining as head of government; he died and was succeeded by Ben Jones.
*1990* Nicholas Braithwaite of the National Democratic Congress (NDC) became prime minister.
*1991* Integration into Windward Islands confederation proposed.

# Guatemala
Republic of
(*República de Guatemala*)

*area* 108,889 sq km/42,031 sq mi
*capital* Guatemala City
*towns* Quezaltenango, Puerto Barrios
*physical* mountainous; narrow coastal plains; limestone tropical plateau in N; frequent earthquakes
*environment* between 1960 and 1980 nearly 57% of the country's forest was cleared for farming
*head of state and government* Jorge Serrano Elías from 1991
*political system* democratic republic
*exports* coffee, bananas, cotton, sugar, beef
*currency* quetzal
*population* (1990 est) 9,340,000 (Mayaquiche Indians 54%, mestizos (mixed race) 42%); growth rate 2.8% p.a. (87% of under-fives suffer from malnutrition)
*life expectancy* men 57, women 61
*languages* Spanish (official); 40% speak 18 Indian languages
*religion* Roman Catholic 80%, Protestant 20%
*literacy* men 63%, women 47% (1985 est)
*GDP* $7 bn (1987); $834 per head

*chronology*

*1839* Independence achieved from Spain.
*1954* Col Carlos Castillo became president in US-backed coup, halting land reform.
*1963* Military coup made Col Enrique Peralta president.
*1966* Cesar Méndez elected president.
*1970* Carlos Araña elected president.
*1974* General Kjell Laugerud became president. Widespread political violence precipitated by the discovery of falsified election returns in March.
*1978* General Fernando Romeo became president.
*1981* Growth of antigovernment guerrilla movement.
*1982* General Angel Anibal became president. Army coup installed General Ríos Montt as head of junta and then as president; political violence continued.
*1983* Montt removed in coup led by General Mejía Victores, who declared amnesty for the guerrillas.
*1985* New constitution adopted; Guatemalan Christian Democratic Party (PDCG) won congressional elections; Vinicio Cerezo elected president.
*1989* Coup attempt against Cerezo foiled. Over 100,000 people killed, and 40,000 reported missing since 1980.
*1991* Jorge Serrano Elías of the Solidarity Action Movement elected president. Diplomatic relations with Belize established.

# Guinea
Republic of
(*République de Guinée*)

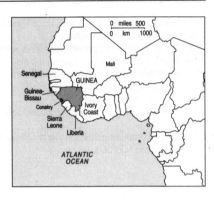

***area*** 245,857 sq km/94,901 sq mi
***capital*** Conakry
***towns*** Labé, Nzérékoré, Kankan
***physical*** flat coastal plain with mountainous interior; sources of rivers Niger, Gambia, and Senegal; forest in SE
***environment*** large amounts of toxic waste from industrialized countries have been dumped in Guinea
***head of state and government*** Lansana Conté from 1984
***political system*** military republic
***exports*** coffee, rice, palm kernels, alumina, bauxite, diamonds
***currency*** syli or franc
***population*** (1990 est) 7,269,000 (chief peoples are Fulani, Malinke, Susu); growth rate 2.3% p.a.
***life expectancy*** men 39, women 42
***languages*** French (official), African languages
***religions*** Muslim 85%, Christian 10%, local 5%
***literacy*** men 40%, women 17% (1985 est)
***GNP*** $1.9 bn (1987); $369 per head
***chronology***
*1958* Full independence achieved from France; Sékou Touré elected president.

*1977* Strong opposition to Touré's rigid Marxist policies forced him to accept return to mixed economy.
*1980* Touré returned unopposed for fourth seven-year term.
*1984* Touré died. Bloodless coup established a military committee for national recovery, led by Col Lansana Conté.
*1985* Attempted coup against Conté while he was out of the country was foiled by loyal troops.
*1990* Sent troops to join the multinational force that attempted to stabilize Liberia.
*1991* Antigovernment general strike by National Confederation of Guinea Workers (CNTG).

# Guinea-Bissau
Republic of
(*República da Guiné-Bissau*)

*area* 36,125 sq km/13,944 sq mi
*capital* Bissau
*towns* Mansôa, São Domingos
*physical* flat coastal plain rising to savanna in E
*head of state and government* João Bernardo Vieira from 1980
*political system* socialist pluralist republic
*exports* rice, coconuts, peanuts, fish, timber
*currency* peso
*population* (1989 est) 929,000; growth rate 2.4% p.a.
*life expectancy* 42; 1990 infant mortality rate was 14.8%
*languages* Portuguese (official), Crioulo (Cape Verdean dialect of Portuguese), African languages
*religions* animism 54%, Muslim 38%, Christian 8%
*literacy* men 46%, women 17% (1985 est)
*GDP* $135 million (1987); $146 per head
*chronology*
*1956* African Party for the Independence of Portuguese Guinea and Cape Verde (PAIGC) formed to secure independence from Portugal.
*1973* Two-thirds of the country declared independent, with Luiz Cabral as president of a state council.
*1974* Independence achieved from Portugal.

*1980* Cape Verde decided not to join a unified state. Cabral deposed, and João Vieira became chair of a council of revolution.
*1981* PAIGC confirmed as the only legal party, with Vieira as its secretary general.
*1982* Normal relations with Cape Verde restored.
*1984* New constitution adopted, making Vieira head of government as well as head of state.
*1989* Vieira re-elected.
*1991* Other parties legalized.
*1992* Multiparty electoral commission established.

## Guyana
Cooperative Republic of

***area*** 214,969 sq km/82,978 sq mi
***capital*** (and port) Georgetown
***towns*** New Amsterdam, Mabaruma
***physical*** coastal plain rises into rolling highlands with savanna in S; mostly tropical rainforest
***head of state*** Desmond Hoyte from 1985
***head of government*** Cheddi Jagan from 1992
***political system*** democratic republic
***exports*** sugar, rice, rum, timber, diamonds, bauxite, shrimps, molasses
***currency*** Guyanese dollar
***population*** (1989 est) 846,000 (51% descendants of workers introduced from India to work the sugar plantations after the abolition of slavery, 30% black, 5% Amerindian); growth rate 2% p.a.
***life expectancy*** men 66, women 71
***languages*** English (official), Hindi, Amerindian
***religions*** Christian 57%, Hindu 33%, Sunni Muslim 9%
***literacy*** men 97%, women 95% (1985 est)
***GNP*** $359 million (1987); $445 per head

## chronology

**1831** Became British colony under name of British Guiana.
**1953** Assembly elections won by left-wing People's Progressive Party (PPP); Britain suspended constitution and installed interim administration, fearing communist takeover.
**1961** Internal self-government granted; Cheddi Jagan became prime minister.
**1964** People's National Congress (PNC) leader Forbes Burnham led PPP–PNC coalition.
**1966** Independence achieved from Britain.
**1970** Guyana became a republic within the Commonwealth.
**1981** Forbes Burnham became first executive president under new constitution.
**1985** Burnham died; succeeded by Desmond Hoyte.
**1992** PPP had decisive victory in assembly elections; Jagan returned as prime minister.

## Haiti
Republic of
(*République d'Haïti*)

*area* 27,750 sq km/10,712 sq mi
*capital* Port-au-Prince
*towns* Cap-Haïtien, Gonaïves, Les Cayes
*physical* mainly mountainous and tropical; occupies W third of Hispaniola Island in Caribbean Sea; seriously deforested
*interim head of state* Joseph Nerette from 1991
*head of government* Marc Bazin from 1992
*political system* transitional
*exports* coffee, sugar, sisal, cotton, cocoa, bauxite
*currency* gourde
*population* (1990 est) 6,409,000; growth rate 1.7% p.a.; one of highest population densities in the world; about 1.5 million Haitians live outside Haiti (in USA and Canada); about 400,000 live in virtual slavery in the Dominican Republic, where they went or were sent to cut sugar cane
*life expectancy* men 51, women 54
*languages* French (official, spoken by literate 10% minority), Creole (spoken by 90% black majority)
*religion* Christian 95% (of which 80% Roman Catholic), voodoo 4%

*literacy* men 40%, women 35% (1985 est)
*GDP* $2.2 bn (1987); $414 per head
*chronology*
*1804* Independence achieved from France.
*1915* Haiti invaded by USA; remained under US control until 1934.
*1957* Dr François Duvalier (Papa Doc) elected president.
*1964* Duvalier pronounced himself president for life.
*1971* Duvalier died, succeeded by his son, Jean-Claude (Baby Doc); thousands murdered during Duvalier era.
*1986* Duvalier deposed; replaced by Lt-Gen Henri Namphy as head of a governing council.
*1988* Feb: Leslie Manigat became president. Namphy staged a military coup in June, but another coup in Sept led by Brig-Gen Prosper Avril replaced him with a civilian government under military control.
*1989* Coup attempt against Avril foiled; US aid resumed.
*1990* Opposition elements expelled; Ertha Pascal-Trouillot acting president.
*1991* Jean-Bertrand Aristide elected president but later overthrown in military coup led by Brig-Gen Raoul Cedras. Efforts to reinstate Aristide failed. Joseph Nerette became interim head of state.
*1992* Economic sanctions imposed since 1991 were eased by the USA but increased by the Organization of American States (OAS). Marc Bazin appointed premier.

**Honduras**
Republic of
(*República de Honduras*)

*area* 112,100 sq km/43,282 sq mi
*capital* Tegucigalpa
*towns* San Pedro Sula; ports La Ceiba, Puerto Cortés
*physical* narrow tropical coastal plain with mountainous interior, Bay Islands
*head of state and government* Rafael Leonardo Callejas from 1990
*political system* democratic republic
*exports* coffee, bananas, meat, sugar, timber (including mahogany, rosewood)
*currency* lempira
*population* (1989 est) 5,106,000 (mestizo, or mixed, 90%, Indians and Europeans 10%); growth rate 3.1% p.a.
*life expectancy* men 58, women 62
*languages* Spanish (official), English, Indian languages
*religion* Roman Catholic 97%
*literacy* men 61%, women 58% (1985 est)
*GDP* $3.5 bn (1987); $758 per head
*chronology*
*1838* Independence achieved from Spain.

*1980* After more than a century of mostly military rule, a civilian government was elected, with Dr Roberto Suazo as president; the commander in chief of the army, General Gustavo Alvarez, retained considerable power.

*1983* Close involvement with the USA in providing naval and air bases and allowing Nicaraguan counter-revolutionaries ('Contras') to operate from Honduras.

*1984* Alvarez ousted in coup led by junior officers, resulting in policy review towards USA and Nicaragua.

*1985* José Azcona elected president after electoral law changed, making Suazo ineligible for presidency.

*1989* Government and opposition declared support for Central American peace plan to demobilize Nicaraguan Contras based in Honduras; Contras and their dependents in Honduras in 1989 thought to number about 55,000.

*1990* Rafael Callejas, National Party (PN), inaugurated as president.

*1992* Border dispute with El Salvador dating from 1861 finally resolved.

# Hungary
Republic of
(*Magyar Köztársaság*)

*area* 93,032 sq km/35,910 sq mi
*capital* Budapest
*towns* Miskolc, Debrecen, Szeged, Pécs
*physical* Great Hungarian Plain covers E half of country; Bakony Forest, Lake Balaton, and Transdanubian Highlands in the W; rivers Danube, Tisza, and Raba
*environment* an estimated 35%–40% of the population live in areas with officially 'inadmissible' air and water pollution.
*head of state* Arpád Göncz from 1990
*head of government* József Antall from 1990
*political system* emergent democratic republic
*exports* machinery, vehicles, iron and steel, chemicals, fruit and vegetables
*currency* forint
*population* (1990 est) 10,546,000 (Magyar 92%, Romany 3%, German 2.5%; Hungarian minority in Romania has caused some friction between the two countries); growth rate 0.2% p.a.
*life expectancy* men 67, women 74

*language* Hungarian (or Magyar), one of the few languages of Europe with non-Indo-European origins
*religions* Roman Catholic 67%, other Christian denominations 25%
*literacy* men 99.3%, women 98.5% (1980)
*GDP* $26.1 bn (1987); $2,455 per head
*chronology*
*1918* Independence achieved from Austro-Hungarian empire.
*1919* A communist state formed for 133 days.
*1920–44* Regency formed under Admiral Horthy, who joined Hitler's attack on the USSR.
*1945* Liberated by USSR.
*1946* Republic proclaimed; Stalinist regime imposed.
*1949* Soviet-style constitution adopted.
*1956* Hungarian national uprising; workers' demonstrations in Budapest; democratization reforms by Imre Nagy overturned by Soviet tanks, János Kádár installed as party leader.
*1968* Economic decentralization reforms.
*1983* Competition introduced into elections.
*1987* VAT and income tax introduced.
*1988* Kádár replaced by Károly Grosz. First free trade union recognized; rival political parties legalized.
*1989* May: border with Austria opened. July: new four-person collective leadership of the Hungarian Socialist Workers' Party (HSWP). Oct: new 'transitional constitution' adopted, founded on multiparty democracy and new presidentialist executive. HSWP changed name to Hungarian Socialist Party (HSP), with Nyers as new leader. Kádár 'retired'.
*1990* HSP reputation damaged by 'Danubegate' bugging scandal. March–April: elections won by right-of-centre coalition, headed by Hungarian Democratic Forum (MDF). May: József Antall, leader of the MDF, appointed premier. Aug: Arpád Göncz elected president.
*1991* Jan: devaluation of currency. June: legislation approved to compensate owners of land and property expropriated under communist government. Last Soviet troops departed. Dec: EC association pact signed.
*1992* March: EC pact came into effect.

## Iceland
Republic of
(*Ly´dveldid Ísland*)

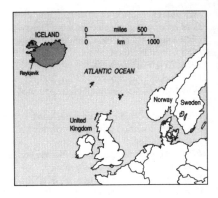

*area* 103,000 sq km/39,758 sq mi
*capital* Reykjavík
*towns* Akureyri, Akranes
*physical* warmed by the Gulf Stream; glaciers and lava fields cover 75% of the country; active volcanoes (Hekla was once thought the gateway to Hell), geysers, hot springs, and new islands created offshore (Surtsey in 1963); subterranean hot water heats 85% of Iceland's homes
*head of state* Vigdís Finnbogadóttir from 1980
*head of government* Davíd Oddsson from 1991
*political system* democratic republic
*exports* cod and other fish products, aluminium, diatomite
*currency* krona
*population* (1990 est) 251,000; growth rate 0.8% p.a.
*life expectancy* men 74, women 80
*language* Icelandic, the most archaic Scandinavian language, in which some of the finest sagas were written
*religion* Evangelical Lutheran 95%
*literacy* 99.9% (1984)
*GDP* $3.9 bn (1986); $16,200 per head

*chronology*
**1944** Independence achieved from Denmark.
**1949** Joined NATO and Council of Europe.
**1953** Joined Nordic Council.
**1976** 'Cod War' with UK.
**1979** Iceland announced 320-km/200-mi exclusive fishing zone.
**1983** Steingrímur Hermannsson appointed to lead a coalition government.
**1985** Iceland declared itself a nuclear-free zone.
**1987** New coalition government formed by Thorsteinn Pálsson after general election.
**1988** Vigdís Finnbogadóttir re-elected president for a third term; Hermannsson led new coalition.
**1991** Davíd Oddsson led new IP–SDP (Independence Party and Social Democratic Party) centre-right coalition, becoming prime minister in the general election.
**1992** Iceland defied world ban to resume whaling industry.

## India
Republic of
(Hindi *Bharat*)

*area* 3,166,829 sq km/1,222,396 sq mi
*capital* Delhi
*towns* Bangalore, Hyderabad, Ahmedabad, Calcutta, Bombay, Madras
*physical* Himalaya mountains on N border; plains around rivers Ganges, Indus, Brahmaputra; Andaman and Nicobar Islands, Lakshadweep (Laccadive Islands)
*environment* the Narmada Valley Project is the world's largest hydroelectric/irrigation scheme. In addition to displacing a million people, the damming of the holy Narmada River will submerge large areas of forest and farmland and create problems of salinization
*head of state* Shankar Dayal Sharma from 1992
*head of government* P V Narasimha Rao from 1991
*political system* liberal democratic federal republic
*exports* tea (world's largest producer), coffee, fish, iron and steel, leather, textiles, clothing, polished diamonds
*currency* rupee
*population* (1991 est) 844,000,000 (920 women to every 1,000 men); growth rate 2.0% p.a.

*life expectancy* men 56, women 55
*languages* Hindi, English, and 14 other official languages: Assamese, Bengali, Gujarati, Kannada, Kashmiri, Malayalam, Marathi, Oriya, Punjabi, Sanskrit, Sindhi, Tamil, Telugu, Urdu
*religions* Hindu 80%, Sunni Muslim 10%, Christian 2.5%, Sikh 2%
*literacy* men 57%, women 29% (1985 est)
*GDP* $220.8 bn (1987); $283 per head
*chronology*
*1947* Independence achieved from Britain.
*1950* Federal republic proclaimed.
*1964* Death of Nehru. Border war with Pakistan over Kashmir.
*1966* Indira Gandhi became prime minister.
*1971* War with Pakistan leading to creation of Bangladesh.
*1975–77* State of emergency proclaimed.
*1977–79* Janata Party government in power.
*1980* Indira Gandhi returned in landslide victory.
*1984* Indira Gandhi assassinated; her son Rajiv Gandhi elected with record majority.
*1987* Public revelation of Bofors corruption scandal.
*1988* New opposition party, Janata Dal, established by former finance minister V P Singh. Voting age lowered from 21 to 18.
*1989* Congress (I) lost majority in general election; Janata Dal minority government formed, with V P Singh prime minister.
*1990* Central rule imposed in Jammu and Kashmir. V P Singh resigned; new minority Janata Dal government formed by Chandra Shekhar. Interethnic and religious violence in Punjab and elsewhere.
*1991* Central rule imposed in Tamil Nadu. Shekhar resigned; elections called for May. May: Rajiv Gandhi assassinated. June: elections resumed, resulting in a Congress (I) minority government led by P V Narasimha Rao. Separatist violence continued.
*1992* Congress (I) won control of state assembly and a majority in parliament in Punjab state elections. Split in Janata Dal opposition resulted in creation of National Front coalition party. Widespread communal violence killed over 1,200 people, mainly Muslims, following destruction of a mosque in Ayodhya by Hindu extremists.
*1993* Sectarian violence in Bombay left 500 dead.

## Indonesia
Republic of
(*Republik Indonesia*)

*area* 1,919,443 sq km/740,905 sq mi
*capital* Jakarta
*towns* Bandung, Surabaya, Semarang, Tandjungpriok
*physical* comprises 13,677 tropical islands, of the Greater Sunda group (including Java and Madura, part of Borneo (Kalimantan), Sumatra, Sulawesi and Belitung), and the Lesser Sundas/Nusa Tenggara (including Bali, Lombok, Sumbawa, Sumba, Flores, and Timor), as well as Malaku/Moluccas and part of New Guinea (Irian Jaya)
*environment* comparison of primary forest and 30-year-old secondary forest has shown that logging in Kalimantan has led to a 20% decline in tree species
*head of state and government* T N J Suharto from 1967
*political system* authoritarian nationalist republic
*exports* coffee, rubber, timber, palm oil, coconuts, tin, tea, tobacco, oil, liquid natural gas
*currency* rupiah
*population* (1989 est) 187,726,000 (including 300 ethnic groups); growth rate 2% p.a.; Indonesia has the world's largest Muslim population; Java is one of the world's most densely populated areas

*life expectancy* men 52, women 55
*languages* Indonesian (official), closely allied to Malay; Javanese is the most widely spoken local dialect
*religions* Muslim 88%, Christian 10%, Buddhist and Hindu 2%
*literacy* men 83%, women 65% (1985 est)
*GDP* $69.7 bn (1987); $409 per head

*chronology*
*1942* Occupied by Japan; nationalist government established.
*1945* Japanese surrender; nationalists declared independence under Achmed Sukarno.
*1949* Formal transfer of Dutch sovereignty.
*1950* Unitary constitution established.
*1963* Western New Guinea (Irian Jaya) ceded by the Netherlands.
*1965–66* Attempted communist coup; General T N J Suharto imposed emergency administration, carried out massacre of hundreds of thousands.
*1967* Sukarno replaced as president by Suharto.
*1975* Guerrillas seeking independence for S Moluccas seized train and Indonesian consulate in the Netherlands, held Western hostages.
*1976* Forced annexation of former Portuguese colony of East Timor.
*1986* Institution of 'transmigration programme' to settle large numbers of Javanese on sparsely populated outer islands, particularly Irian Jaya.
*1988* Partial easing of travel restrictions to East Timor. Suharto re-elected for fifth term.
*1989* Foreign debt reaches $50 billion; Western creditors offer aid on condition that concessions are made to foreign companies and that austerity measures are introduced.
*1991* Democracy forums launched to promote political dialogue. Massacre in East Timor.
*1992* The ruling Golkar party won the assembly elections.
*1993* President Suharto re-elected for sixth consecutive five-year term.

# Iran

Islamic Republic of
(*Jomhori-e-Islami-e-Irân*; until 1935 *Persia*)

*area* 1,648,000 sq km/636,128 sq mi
*capital* Tehran
*towns* Isfahan, Mashhad, Tabriz, Shiraz, Ahvaz, Abadan
*physical* plateau surrounded by mountains, including Elburz and Zagros; Lake Rezayeh; Dasht-Ekavir Desert; occupies islands of Abu Musa, Greater Tunb and Lesser Tunb in the Gulf
*Leader of the Islamic Revolution* Seyed Ali Khamenei from 1989
*head of government* Ali Akbar Hoshemi Rafsanjani from 1989
*political system* authoritarian Islamic republic
*exports* carpets, cotton textiles, metalwork, leather goods, oil, petrochemicals, fruit
*currency* rial
*population* (1989 est) 51,005,000 (including minorities in Azerbaijan, Baluchistan, Khuzestan/Arabistan, and Kurdistan); growth rate 3.2% p.a.
*life expectancy* men 57, women 57
*languages* Farsi (official), Kurdish, Turkish, Arabic, English, French
*religion* Shi'ite Muslim (official) 92%, Sunni Muslim 5%, Zoroastrian 2%, Jewish, Baha'i, and Christian 1%

*literacy* men 62%, women 39% (1985 est)
*GDP* $86.4 bn (1987); $1,756 per head
*chronology*
*1946* British, US, and Soviet forces left Iran.
*1951* Oilfields nationalized by Prime Minister Muhammad Mossadeq.
*1953* Mossadeq deposed and the US-backed shah, Muhammed Reza Shah Pahlavi, took full control of the government.
*1975* The shah introduced single-party system.
*1978* Opposition to the shah organized from France by Ayatollah Khomeini.
*1979* Shah left the country; Khomeini returned to create Islamic state. Revolutionaries seized US hostages at embassy in Tehran; US economic boycott.
*1980* Start of Iran–Iraq War.
*1981* US hostages released.
*1984* Egyptian peace proposals rejected.
*1985* Fighting intensified in Iran–Iraq War.
*1988* Cease-fire; talks with Iraq began.
*1989* Khomeini called for the death of British writer Salman Rushdie. June: Khomeini died; Ali Khamenei elected interim Leader of the Revolution; speaker of Iranian parliament Hoshemi Rafsanjani elected president. Secret oil deal with Israel revealed.
*1990* Generous peace terms with Iraq accepted. Normal relations with UK restored.
*1991* Imprisoned British business executive released. Nearly one million Kurds arrived in Iran from Iraq, fleeing persecution by Saddam Hussein after the Gulf War.
*1992* Pro-Rafsanjani moderates won assembly elections.

# Iraq
Republic of
(*al Jumhouriya al 'Iraqia*)

*area* 434,924 sq km/167,881 sq mi
*capital* Baghdad
*towns* Mosul, Basra
*physical* mountains in N, desert in W; wide valley of rivers Tigris and Euphrates NW–SE
*environment* a chemical-weapons plant covering an area of 65 sq km/25 sq mi, situated 80 km/50 mi NW of Baghdad, has been described by the UN as the largest toxic waste dump in the world
*head of state and government* Saddam Hussein al-Tikriti from 1979
*political system* one-party socialist republic
*exports* oil (prior to UN sanctions), wool, dates (80% of world supply)
*currency* Iraqi dinar
*population* (1989 est) 17,610,000 (Arabs 77%, Kurds 19%, Turks 2%); growth rate 3.6% p.a.
*life expectancy* men 62, women 63
*languages* Arabic (official); Kurdish, Assyrian, Armenian
*religions* Shi'ite Muslim 60%, Sunni Muslim 37%, Christian 3%
*literacy* men 68%, women 32% (1980 est)
*GDP* $42.3 bn (1987); $3,000 per head

*chronology*
*1920* Iraq became a British League of Nations protectorate.
*1921* Hashemite dynasty established, with Faisal I installed by Britain as king.
*1932* Independence achieved from British protectorate status.
*1958* Monarchy overthrown; Iraq became a republic.
*1963* Joint Ba'athist-military coup headed by Col Salem Aref.
*1968* Military coup put Maj-Gen al-Bakr in power.
*1979* Al-Bakr replaced by Saddam Hussein.
*1980* War between Iraq and Iran broke out.
*1985* Fighting intensified.
*1988* Cease-fire; talks began with Iran. Iraq used chemical weapons against Kurdish rebels seeking greater autonomy.
*1989* Unsuccessful coup against President Hussein; Iraq launched ballistic missile in successful test.
*1990* Peace treaty favouring Iran agreed. Aug: Iraq invaded and annexed Kuwait, precipitating another Gulf crisis. US forces massed in Saudi Arabia at request of King Fahd. United Nations resolutions ordered Iraqi withdrawal from Kuwait and imposed total trade ban on Iraq; UN resolution sanctioning force approved. All foreign hostages released.
*1991* 16 Jan: US-led forces began aerial assault on Iraq; Iraq's infrastructure destroyed by bombing. 23–28 Feb: land–sea–air offensive to free Kuwait successful. Uprisings of Kurds and Shi'ites brutally suppressed by surviving Iraqi troops. Talks between Kurdish leaders and Saddam Hussein about Kurdish autonomy. Allied troops withdrew after establishing 'safe havens' for Kurds in the north, leaving a rapid-reaction force near the Turkish border. Allies threatened to bomb strategic targets in Iraq if full information about nuclear facilities denied to UN.
*1992* UN imposed a 'no-fly zone' over S Iraq to protect Shi'ites.
*1993* Jan: Iraqi incursions into the 'no-fly zone' prompted US-led alliance aircraft to bomb 'strategic' targets in Iraq. Relations subsequently improved.

# Ireland
Republic of
(*Eire*)

*area* 70,282 sq km/27,146 sq mi
*capital* Dublin
*towns* Cork, Dun Laoghaire, Limerick, Waterford
*physical* central plateau surrounded by hills; rivers Shannon, Liffey, Boyne
*head of state* Mary Robinson from 1990
*head of government* Albert Reynolds from 1992
*political system* democratic republic
*exports* livestock, dairy products, Irish whiskey, microelectronic components and assemblies, mining and engineering products, chemicals, clothing; tourism is important
*currency* punt
*population* (1989 est) 3,734,000; growth rate 0.1% p.a.
*life expectancy* men 70, women 76
*languages* Irish Gaelic and English (both official)
*religion* Roman Catholic 94%
*literacy* 99% (1984)
*GDP* $21.9 (1987); $6,184 per head

*chronology*
*1937* Independence achieved from Britain.
*1949* Eire left the Commonwealth and became the Republic of Ireland.
*1973* Fianna Fáil defeated after 40 years in office; Liam Cosgrave formed a coalition government.
*1977* Fianna Fáil returned to power, with Jack Lynch as prime minister.
*1979* Lynch resigned, succeeded by Charles Haughey.
*1981* Garret FitzGerald formed a coalition.
*1983* New Ireland Forum formed, but rejected by the British government.
*1985* Anglo-Irish Agreement signed.
*1986* Protests by Ulster Unionists against the agreement.
*1987* General election won by Charles Haughey.
*1988* Relations with UK at low ebb because of disagreement over extradition decisions.
*1989* Haughey failed to win majority in general election. Progressive Democrats given cabinet positions in coalition government.
*1990* Mary Robinson elected president; John Bruton became Fine Gael leader.
*1992* Jan: Haughey resigned after losing parliamentary majority. Feb: Albert Reynolds became Fianna Fáil leader and prime minister. June: National referendum approved ratification of Maastricht Treaty. Nov: Reynolds lost confidence vote; election result inconclusive.
*1993* Fianna Fáil–Labour coalition formed.

# Israel
State of
(*Medinat Israel*)

*area* 20,800 sq km/8,029 sq mi (as at 1949 armistice)
*capital* Jerusalem (not recognized by the United Nations)
*towns* ports Tel Aviv/Jaffa, Haifa, Acre, Eilat, Holon, Beersheba
*physical* coastal plain of Sharon between Haifa and Tel Aviv noted since ancient times for fertility; central mountains of Galilee, Samariq, and Judea; Dead Sea, Lake Tiberias, and river Jordan Rift Valley along the E are below sea level; Negev Desert in the S; Israel occupies Golan Heights, West Bank, and Gaza
*head of state* Ezer Weizman from 1993
*head of government* Yitzhak Rabin from 1992
*political system* democratic republic
*exports* citrus and other fruit, fertilizers, diamonds, plastics, petrochemicals, textiles, electronics (military, medical, scientific, industrial), precision instruments, aircraft and missiles
*currency* shekel
*population* (1989 est) 4,477,000 (including 750,000 Arab Israeli citizens and over 1 million Arabs in the occupied territories); under the Law of Return 1950, 'every Jew shall be entitled to come to Israel as an immigrant'; those from the East and E Europe are Ashkenazim, and

those from Mediterranean Europe (Spain, Portugal, Italy, France, Greece) and Arab N Africa are Sephardim (over 50% of the population is now of Sephardic descent). Between Jan 1990 and April 1991, 250,000 Soviet Jews emigrated to Israel. An Israeli-born Jew is a Sabra. About 500,000 Israeli Jews are resident in the USA. Growth rate 1.8% p.a.
*life expectancy* men 73, women 76
*languages* Hebrew and Arabic (official); Yiddish, European and W Asian languages
*religions* Israel is a secular state, but the predominant faith is Judaism 83%; also Sunni Muslim, Christian, and Druse
*literacy* Jewish 88%, Arab 70%
*GDP* $35 bn (1987); $8,011 per head
*chronology*
*1948* Independent State of Israel proclaimed with David Ben-Gurion as prime minister; attacked by Arab nations, Israel won the War of Independence. Many displaced Arabs settled in refugee camps in the Gaza Strip and West Bank.
*1952* Col Gamal Nasser of Egypt stepped up blockade of Israeli ports and support of Arab guerrillas in Gaza.
*1956* Israel invaded Gaza and Sinai.
*1959* Egypt renewed blockade of Israeli trade through Suez Canal.
*1963* Ben-Gurion resigned, succeeded by Levi Eshkol.
*1964* Palestine Liberation Organization (PLO) founded with the aim of overthrowing the state of Israel.
*1967* Israel victorious in the Six-Day War. Gaza, West Bank, E Jerusalem, Sinai, and Golan Heights captured.
*1968* Israel Labour Party formed, led by Golda Meir.
*1969* Golda Meir became prime minister.
*1973* Yom Kippur War: Israel attacked by Egypt and Syria.
*1974* Golda Meir succeeded by Yitzhak Rabin.
*1975* Suez Canal reopened.
*1977* Menachem Begin elected prime minister. Egyptian president addressed the Knesset.
*1978* Camp David talks.
*1979* Egyptian–Israeli agreement signed. Israel agreed to withdraw

from Sinai.
*1980* Jerusalem declared capital of Israel.
*1981* Golan Heights formally annexed.
*1982* Israel pursued PLO fighters into Lebanon.
*1983* Peace treaty between Israel and Lebanon signed but not ratified.
*1985* Formation of government of national unity with Labour and Likud ministers.
*1986* Yitzhak Shamir took over from Peres under power-sharing agreement.
*1987* Outbreak of Palestinian uprising (Intifada) in West Bank and Gaza.
*1988* Criticism of Israel's handling of Palestinian uprising in occupied territories; PLO acknowledged Israel's right to exist.
*1989* New Likud–Labour coalition government formed under Shamir. Limited progress achieved on proposals for negotiations leading to elections in occupied territories.
*1990* Coalition collapsed due to differences over peace process. New Shamir right-wing coalition formed.
*1991* Shamir gave cautious response to Middle East peace proposals. Some Palestinian prisoners released. Peace talks began in Madrid.
*1992* Jan: Shamir lost majority in Knesset when ultra-Orthodox party withdrew from coalition. June: Labour Party, led by Yitzhak Rabin, won elections; coalition formed under Rabin. Aug: US–Israeli loan agreement signed. Dec: Palestinians expelled in face of international criticism.
*1993* Jan: UN condemned expulsion of Palestinians. Ban on contacts with PLO formally lifted. Feb: government allowed 100 of 400 expelled Palestinians to return. March: Ezer Weizman elected president; Binyamin 'Bibi' Netanyahu elected leader of Likud party.

# Italy
Republic of
(*Repubblica Italiana*)

***area*** 301,300 sq km/116,332 sq mi
***capital*** Rome
***towns*** Milan, Turin, Naples, Genoa, Palermo, Trieste
***physical*** mountainous (Maritime Alps, Dolomites, Apennines) with narrow coastal lowlands; rivers Po, Adige, Arno, Tiber, Rubicon; islands of Sicily, Sardinia, Elba, Capri, Ischia, Lipari, Pantelleria; lakes Como, Maggiore, Garda
***environment*** Milan has the highest recorded level of sulphur-dioxide pollution of any city in the world. The Po River, with pollution ten times higher than officially recommended levels, is estimated to discharge around 250 tonnes of arsenic into the sea each year
***head of state*** Oscar Luigi Scalfaro from 1992
***head of government*** Carlo Azeglio Ciampi from 1993
***political system*** democratic republic
***exports*** wine (world's largest producer), fruit, vegetables, textiles (Europe's largest silk producer), clothing, leather goods, motor vehicles, electrical goods, chemicals, marble (Carrara), sulphur, mercury, iron, steel
***currency*** lira

*population* (1990 est) 57,657,000; growth rate 0.1% p.a.
*life expectancy* men 73, women 80 (1989)
*language* Italian; German, French, Slovene, and Albanian minorities
*religion* Roman Catholic 100% (state religion)
*literacy* 97% (1989)
*GDP* $748 bn; $13,052 per head (1988)
*chronology*
*1946* Monarchy replaced by a republic.
*1948* New constitution adopted.
*1954* Trieste returned to Italy.
*1976* Communists proposed establishment of broad-based, left–right government, the 'historic compromise'; rejected by Christian Democrats.
*1978* Christian Democrat Aldo Moro, architect of the historic compromise, kidnapped and murdered by Red Brigade guerrillas.
*1983* Bettino Craxi, a Socialist, became leader of broad coalition government.
*1987* Craxi resigned; succeeding coalition fell within months.
*1988* Christian Democrats' leader Ciriaco de Mita established a five-party coalition including the Socialists.
*1989* De Mita resigned after disagreements within his coalition government; succeeded by Giulio Andreotti. De Mita lost leadership of Christian Democrats; Communists formed 'shadow government'.
*1991* Referendum approved electoral reform.
*1992* April: ruling coalition lost its majority in general election; President Cossiga resigned, replaced by Oscar Luigi Scalfaro in May. Giuliano Amato, deputy leader of Democratic Party of the Left (PDS), accepted premiership. Sept: lira devalued and Italy withdrew from the Exchange Rate Mechanism.
*1993* Feb–March: investigation of corruption network exposed Mafia links with several notable politicians, including Craxi and Andreotti. Craxi resigned Socialist Party leadership; replaced by Giorgio Benvenutu. April: referendum results showed Italian people strongly in favour of majority electoral system. Amato resigned premiership; Carlo Ciampi named as his successor.

# Ivory Coast
Republic of
(*République de la Côte d'Ivoire*)

*area* 322,463 sq km/124,471 sq mi
*capital* Yamoussoukro
*towns* Bouaké, Daloa, Man, Abidjan, San-Pédro
*physical* tropical rainforest (diminishing as exploited) in S; savanna and low mountains in N
*environment* an estimated 85% of the country's forest has been destroyed by humans
*head of state and government* Félix Houphouët-Boigny from 1960
*political system* emergent democratic republic
*exports* coffee, cocoa, timber, petroleum products
*currency* franc CFA
*population* (1990 est) 12,070,000; growth rate 3.3% p.a.
*life expectancy* men 52, women 55 (1989)
*languages* French (official), over 60 native dialects
*religions* animist 65%, Muslim 24%, Christian 11%
*literacy* 35% (1988)
*GDP* $7.6 bn (1987); $687 per head
*chronology*
*1904* Became part of French West Africa.

*1958* Achieved internal self-government.
*1960* Independence achieved from France, with Félix Houphouët-Boigny as president of a one-party state.
*1985* Houphouët-Boigny re-elected, unopposed.
*1986* Name changed officially from Ivory Coast to Côte d'Ivoire.
*1990* Houphouët-Boigny and Democratic Party of the Ivory Coast (PDCI) re-elected.

# Jamaica

*area* 10,957 sq km/4,230 sq mi
*capital* Kingston
*towns* Montego Bay, Spanish Town, St Andrew
*physical* mountainous tropical island
*head of state* Elizabeth II from 1962, represented by governor general
*head of government* P J Patterson from 1992
*political system* constitutional monarchy
*exports* sugar, bananas, bauxite, rum, cocoa, coconuts, liqueurs, cigars, citrus
*currency* Jamaican dollar
*population* (1990 est) 2,513,000 (African 76%, mixed 15%, Chinese, Caucasian, East Indian); growth rate 2.2% p.a.
*life expectancy* men 75, women 78 (1989)
*languages* English, Jamaican creole
*religions* Protestant 70%, Rastafarian
*literacy* 82% (1988)
*GDP* $2.9 bn; $1,187 per head (1989)
*chronology*
*1494* Columbus reached Jamaica.

*1509–1655* Occupied by Spanish.
*1655* Captured by British.
*1944* Internal self-government introduced.
*1962* Independence achieved from Britain, with Alexander Bustamante of the Jamaica Labour Party (JLP) as prime minister.
*1967* JLP re-elected under Hugh Shearer.
*1972* Michael Manley of the People's National Party (PNP) became prime minister.
*1980* JLP elected, with Edward Seaga as prime minister.
*1983* JLP re-elected, winning all 60 seats.
*1988* Island badly damaged by Hurricane Gilbert.
*1989* PNP won a decisive victory with Michael Manley returning as prime minister.
*1992* Manley resigned, succeeded by P J Patterson.
*1993* March: landslide victory for PNP in general election.

# Japan
(*Nippon*)

*area* 377,535 sq km/145,822 sq mi
*capital* Tokyo
*towns* Fukuoka, Kyoto, Sapporo, Osaka, Nagoya, Yokohama
*physical* mountainous, volcanic; comprises over 1,000 islands, the largest of which are Hokkaido, Honshu, Kyushu, and Shikoku
*head of state* (figurehead) Emperor Akihito from 1989
*head of government* Kiichi Miyazawa from 1991
*political system* liberal democracy
*exports* televisions, cassette and video recorders, radios, cameras, computers, robots, other electronic and electrical equipment, motor vehicles, ships, iron, steel, chemicals, textiles
*currency* yen
*population* (1990 est) 123,778,000; growth rate 0.5% p.a.
*life expectancy* men 76, women 82 (1989)
*language* Japanese
*religions* Shinto, Buddhist (often combined), Christian.
*literacy* 99% (1989)
*GDP* $2.4 trillion; $19,464 per head (1989)
*chronology*
*1867* End of shogun rule; executive power passed to emperor. Start of modernization of Japan.

***1894–95*** War with China; Formosa (Taiwan) and S Manchuria gained.
***1902*** Formed alliance with Britain.
***1904–05*** War with Russia; Russia ceded southern half of Sakhalin.
***1910*** Japan annexed Korea.
***1914*** Joined Allies in World War I.
***1918*** Received German Pacific islands as mandates.
***1931–32*** War with China; renewed 1937.
***1941*** Japan attacked US fleet at Pearl Harbor 7 Dec.
***1945*** World War II ended with Japanese surrender. Allied control commission took power. Formosa and Manchuria returned to China.
***1946*** Framing of 'peace constitution'. Emperor Hirohito became figurehead ruler.
***1952*** Full sovereignty regained.
***1958*** Joined United Nations.
***1968*** Bonin and Volcano Islands regained.
***1972*** Ryukyu Islands regained.
***1974*** Prime Minister Tanaka resigned over Lockheed bribes scandal.
***1982*** Yasuhiro Nakasone elected prime minister.
***1985*** Yen revalued.
***1987*** Noboru Takeshita chosen to succeed Nakasone.
***1988*** Recruit scandal cast shadow over government and opposition parties.
***1989*** Emperor Hirohito died; succeeded by his son Akihito. Many cabinet ministers implicated in Recruit scandal and Takeshita resigned; succeeded by Sosuke Uno. Aug: Uno resigned after sex scandal; succeeded by Toshiki Kaifu.
***1990*** Feb: new house of councillors' elections won by the Liberal Democratic Party (LDP). Public-works budget increased by 50% to encourage imports.
***1991*** Japan contributed billions of dollars to the Gulf War and its aftermath. Kaifu succeeded by Kiichi Miyazawa.
***1992*** Over 100 politicians implicated in new financial scandal. Emperor Akihito made first Japanese imperial visit to China. Trade surpluses reached record levels.
***1993*** Worst recession of postwar era; trade surpluses, however, again reached record levels.

# Jordan
Hashemite Kingdom of (*Al Mamlaka al Urduniya al Hashemiyah*)

*area* 89,206 sq km/34,434 sq mi (West Bank 5,879 sq km/2,269 sq mi)
*capital* Amman
*towns* Zarqa, Irbid, Aqaba
*physical* desert plateau in E; rift valley separates E and W banks of the river Jordan
*head of state* King Hussein ibn Talai from 1952
*head of government* Mudar Badran from 1989
*political system* constitutional monarchy
*exports* potash, phosphates, citrus, vegetables
*currency* Jordanian dinar
*population* (1990 est) 3,065,000 (including Palestinian refugees); West Bank (1988) 866,000; growth rate 3.6% p.a.
*life expectancy* men 67, women 71
*languages* Arabic (official), English
*religions* Sunni Muslim 92%, Christian 8%
*literacy* 71% (1988)
*GDP* $4.3 bn (1987); $1,127 per head (1988)
*chronology*
*1946* Independence achieved from Britain as Transjordan.
*1949* New state of Jordan declared.

***1950*** Jordan annexed West Bank.
***1953*** Hussein ibn Talai officially became king of Jordan.
***1958*** Jordan and Iraq formed Arab Federation that ended when the Iraqi monarchy was deposed.
***1967*** Israel captured and occupied West Bank. Martial law imposed.
***1976*** Lower house dissolved, political parties banned, elections postponed until further notice.
***1982*** Hussein tried to mediate in Arab–Israeli conflict.
***1984*** Women voted for the first time.
***1985*** Hussein and Yassir Arafat put forward framework for Middle East peace settlement. Secret meeting between Hussein and Israeli prime minister.
***1988*** Hussein announced decision to cease administering the West Bank as part of Jordan, passing responsibility to Palestine Liberation Organization, and the suspension of parliament.
***1989*** Prime Minister Zaid al-Rifai resigned; Hussein promised new parliamentary elections following criticism of economic policies. Riots over price increases up to 50% following fall in oil revenues. First parliamentary elections for 22 years; Muslim Brotherhood won 25 of 80 seats but exiled from government; martial law lifted.
***1990*** Hussein unsuccessfully tried to mediate after Iraq's invasion of Kuwait. Massive refugee problems as thousands fled to Jordan from Kuwait and Iraq.
***1991*** 24 years of martial law ended; ban on political parties lifted.
***1992*** Political parties allowed to register.

# Kazakhstan
Republic of

*area* 2,717,300 sq km/1,049,150 sq mi
*capital* Alma-Ata
*towns* Karaganda, Semipalatinsk, Petropavlovsk
*physical* Caspian and Aral seas, Lake Balkhash; Steppe region
*head of state* Nursultan Nazarbayev from 1990
*head of government* Sergey Tereshchenko from 1991
*political system* emergent democracy
*products* grain, copper, lead, zinc, manganese, coal, oil
*population* (1990) 16,700,000 (Kazakh 40%, Russian 38%, German 6%, Ukrainian 5%)
*language* Russian; Kazakh, related to Turkish
*religion* Sunni Muslim
*chronology*
*1920* Autonomous republic in USSR.
*1936* Joined the USSR and became a full union republic.
*1950s* Site of Nikita Khrushchev's ambitious 'Virgin Lands' agricultural extension programme.

***1960s*** A large influx of Russian settlers turned the Kazakhs into a minority in their own republic.
***1986*** Riots in Alma-Alta after Gorbachev ousted local communist leader.
***1989*** June: Nursultan Nazarbayev became leader of the Kazakh Communist Party (KCP) and instituted economic and cultural reform programmes.
***1990*** Feb: Nazarbayev became head of state.
***1991*** March: support pledged for continued union with USSR; Aug: Nazarbayev condemned attempted anti-Gorbachev coup; KCP abolished and replaced by Independent Socialist Party of Kazakhstan (SPK). Dec: joined new Commonwealth of Independent States (CIS); independence recognized by USA.
***1992*** Jan: admitted into Conference on Security and Cooperation in Europe (CSCE). March: became a member of the United Nations. May: trade agreement with USA.
***1993*** Jan: new constitution adopted, increasing the authority of the president and making Kazakh the state language.

## Kenya
Republic of
(*Jamhuri ya Kenya*)

*area* 582,600 sq km/224,884 sq mi
*capital* Nairobi
*towns* Kisumu, Mombasa
*physical* mountains and highlands in W and centre; coastal plain in S; arid interior and tropical coast
*environment* the elephant faces extinction as a result of poaching
*head of state and government* Daniel arap Moi from 1978
*political system* authoritarian nationalism
*exports* coffee, tea, pineapples, petroleum products
*currency* Kenya shilling
*population* (1990 est) 25,393,000 (Kikuyu 21%, Luo 13%, Luhya 14%, Kelenjin 11%; Asian, Arab, European); growth rate 4.2% p.a.
*life expectancy* men 59, women 63 (1989)
*languages* Kiswahili (official), English; there are many local dialects
*religions* Protestant 38%, Roman Catholic 28%, indigenous beliefs 26%, Muslim 6%
*literacy* 50% (1988)
*GDP* $6.9 bn (1987); $302 per head (1988)
*chronology*
*1895* British East African protectorate established.
*1920* Kenya became a British colony.

*1944* African participation in politics began.
*1950* Mau Mau campaign began.
*1953* Nationalist leader Jomo Kenyatta imprisoned by British authorities.
*1956* Mau Mau campaign defeated, Kenyatta released.
*1963* Achieved internal self-government, with Kenyatta as prime minister.
*1964* Independence achieved from Britain as a republic within the Commonwealth, with Kenyatta as president.
*1967* East African Community (EAC) formed with Tanzania and Uganda.
*1977* Collapse of EAC.
*1978* Death of Kenyatta. Succeeded by Daniel arap Moi.
*1982* Attempted coup against Moi foiled.
*1983* Moi re-elected unopposed.
*1984* Over 2,000 people massacred by government forces at Wajir.
*1985-86* Thousands of forest villagers evicted and their homes destroyed to make way for cash crops.
*1988* Moi re-elected. 150,000 evicted from state-owned forests.
*1989* Moi announced release of all known political prisoners. Confiscated ivory burned in attempt to stop elephant poaching.
*1990* Despite antigovernment riots, Moi refused multiparty politics.
*1991* Increasing demands for political reform; Moi promised multiparty politics.
*1992* Constitutional amendment passed. Dec: Moi re-elected in first direct elections despite allegations of fraud.

# Kiribati
Republic of

*area* 717 sq km/277 sq mi
*capital* (and port) Bairiki (on Tarawa Atoll)
*physical* comprises 33 Pacific coral islands: the Kiribati (Gilbert), Rawaki (Phoenix), Banaba (Ocean Island), and three of the Line Islands including Kiritimati (Christmas Island)
*environment* the islands are threatened by the possibility of a rise in sea level caused by global warming. A rise of approximately 30 cm/ 1 ft by the year 2040 will make existing fresh water brackish and undrinkable
*head of state and government* Teatao Teannaki from 1991
*political system* liberal democracy
*exports* copra, fish
*currency* Australian dollar
*population* (1990 est) 65,600 (Micronesian); growth rate 1.7% p.a.
*languages* English (official), Gilbertese
*religions* Roman Catholic 48%, Protestant 45%
*literacy* 90% (1985)
*GDP* $26 million (1987); $430 per head (1988)
*chronology*
*1892* Gilbert and Ellice Islands proclaimed a British protectorate.
*1937* Phoenix Islands added to colony.

***1950s*** UK tested nuclear weapons on Kiritimati (formerly Christmas Island).
***1962*** USA tested nuclear weapons on Kiritimati.
***1975*** Ellice Islands separated to become Tuvalu.
***1977*** Gilbert Islands achieved internal self-government.
***1979*** Independence achieved from Britain, within the Commonwealth, as the Republic of Kiribati, with Ieremia Tabai as president.
***1982 and 1983*** Tabai re-elected.
***1985*** Fishing agreement with Soviet state-owned company negotiated, prompting formation of Kiribati's first political party, the opposition Christian Democrats.
***1987*** Tabai re-elected.
***1991*** Tabai re-elected but not allowed under constitution to serve further term; Teatao Teannaki elected president.

# Korea, North
Democratic People's Republic of (*Chosun Minchu-chui Inmin Konghwa-guk*)

*area* 120,538 sq km/46,528 sq mi
*capital* Pyongyang
*towns* Chongjin, Nampo, Wonsan
*physical* wide coastal plain in W rising to mountains cut by deep valleys in interior
*environment* the building of a hydroelectric dam at Kumgangsan on a tributary of the Han River has been opposed by South Korea as a potential flooding threat to central Korea
*head of state* Kim Il Sung from 1972 (also head of Korean Workers' Party)
*head of government* Kang Song San from 1992
*political system* communism
*exports* coal, iron, copper, textiles, chemicals
*currency* won
*population* (1990 est) 23,059,000; growth rate 2.5% p.a.
*life expectancy* men 67, women 73 (1989)
*language* Korean
*religions* traditionally Buddhist, Confucian, but religious activity curtailed by the state
*literacy* 99% (1989)
*GNP* $20 bn; $3,450 per head (1988)

*chronology*
**1910** Korea formally annexed by Japan.
**1945** Russian and US troops entered Korea, forced surrender of Japanese, and divided the country in two. Soviet troops occupied North Korea.
**1948** Democratic People's Republic of Korea declared.
**1950** North Korea invaded South Korea to unite the nation, beginning the Korean War.
**1953** Armistice agreed to end Korean War.
**1961** Friendship and mutual assistance treaty signed with China.
**1972** New constitution, with executive president, adopted. Talks took place with South Korea about possible reunification.
**1980** Reunification talks broke down.
**1983** Four South Korean cabinet ministers assassinated in Rangoon, Burma (Myanmar), by North Korean army officers.
**1985** Increased relations with the USSR.
**1989** Increasing evidence shown of nuclear-weapons development.
**1990** Diplomatic contacts with South Korea and Japan suggested the beginning of a thaw in North Korea's relations with the rest of the world.
**1991** Became a member of the United Nations. Signed nonaggression agreement with South Korea; agreed to ban nuclear weapons.
**1992** Signed Nuclear Safeguards Agreement, allowing international inspection of its nuclear facilities. Also signed a pact with South Korea for mutual inspection of nuclear facilities. Passed legislation making foreign investment in the country attractive. Yon Hyong Muk replaced by Kang Song San.
**1993** March: government announced it was pulling out of Nuclear Non-Proliferation Treaty.

# Korea, South
Republic of Korea
(*Daehan Minguk*)

*area* 98,799 sq km/38,161 sq mi
*capital* Seoul
*towns* Taegu, Pusan, Inchon
*physical* southern end of a mountainous peninsula separating the Sea of Japan from the Yellow Sea
*head of state* Kim Young Sam from 1992
*head of government* Hwang In Sung from 1993
*political system* emergent democracy
*exports* steel, ships, chemicals, electronics, textiles and clothing, plywood, fish
*currency* won
*population* (1990 est) 43,919,000; growth rate 1.4% p.a.
*life expectancy* men 66, women 73 (1989)
*language* Korean
*religions* traditionally Buddhist, Confucian, and Chondokyo; Christian 28%
*literacy* 92% (1989)
*GNP* $171bn (1988); $2,180 per head (1986)
*chronology*
*1910* Korea formally annexed by Japan.
*1945* Russian and US troops entered Korea, forced surrender of

Japanese, and divided the country in two. US military government took control of South Korea.
*1948* Republic proclaimed.
*1950-53* War with North Korea.
*1960* President Syngman Rhee resigned amid unrest.
*1961* Military coup by General Park Chung-Hee. Industrial growth programme.
*1979* Assassination of President Park.
*1980* Military takeover by General Chun Doo Hwan.
*1987* Adoption of more democratic constitution after student unrest. Roh Tae Woo elected president.
*1988* Former president Chun, accused of corruption, publicly apologized and agreed to hand over his financial assets to the state. Seoul hosted Summer Olympic Games.
*1989* Roh reshuffled cabinet, threatened crackdown on protesters.
*1990* Two minor opposition parties united with Democratic Justice Party to form ruling Democratic Liberal Party (DLP). Diplomatic relations established with the USSR.
*1991* Violent mass demonstrations against the government. New opposition grouping, the Democratic Party, formed. Prime Minister Ro Jai Bong replaced by Chung Won Shik. Entered United Nations. Nonaggression and nuclear pacts signed with North Korea.
*1992* DLP lost absolute majority in March general election; substantial gains made by Democratic Party and newly formed Unification National Party (UNP), led by Chung Ju Wong. Diplomatic relations with China established. Dec: Kim Young Sam, DLP candidate, won the presidential election.
*1993* Feb: Kim Young Sam assumed office. Kim Dae Jung and Chung Ju Yung announced their retirement from active politics. Hwang In Sung appointed prime minister.

# Kuwait
State of
(*Dowlat al Kuwait*)

*area* 17,819 sq km/6,878 sq mi
*capital* Kuwait
*towns* Jahra, Ahmadi, Fahaheel
*physical* hot desert; islands of Failaka, Bubiyan, and Warba at NE corner of Arabian Peninsula
*environment* during the Gulf War 1990–91, 650 oil wells were set alight and about 300,000 tonnes of oil were released into the waters of the Gulf leading to pollution haze, photochemical smog, acid rain, soil contamination, and water pollution
*head of state and government* Jabir al-Ahmad al-Jabir al-Sabah from 1977
*political system* absolute monarchy
*exports* oil
*currency* Kuwaiti dinar
*population* (1990 est) 2,080,000 (Kuwaitis 40%, Palestinians 30%); growth rate 5.5% p.a.
*life expectancy* men 72, women 76 (1989)
*languages* Arabic 78%, Kurdish 10%, Farsi 4%
*religion* Sunni Muslim 45%, Shi'ite minority 30%
*literacy* 71% (1988)

***GNP*** $19.1 bn; $10,410 per head (1988)
*chronology*
***1914*** Britain recognized Kuwait as an independent sovereign state.
***1961*** Full independence achieved from Britain, with Sheik Abdullah al-Salem al-Sabah as emir.
***1965*** Sheik Abdullah died; succeeded by his brother, Sheik Sabah.
***1977*** Sheik Sabah died; succeeded by Crown Prince Jabir.
***1983*** Shi'ite guerrillas bombed targets in Kuwait; 17 arrested.
***1986*** National assembly suspended.
***1987*** Kuwaiti oil tankers reflagged, received US Navy protection; missile attacks by Iran.
***1988*** Aircraft hijacked by pro-Iranian Shi'ites demanding release of convicted guerrillas; Kuwait refused.
***1989*** Two of the convicted guerrillas released.
***1990*** Prodemocracy demonstrations suppressed. Kuwait annexed by Iraq. Emir set up government in exile in Saudi Arabia.
***1991*** Feb: Kuwait liberated by US-led coalition forces; extensive damage to property and environment. New government omitted any opposition representatives. Trials of alleged Iraqi collaborators criticized.
***1992*** Oct: reconstituted national assembly elected on restricted franchise, with opposition party winning majority of seats.
***1993*** Jan: incursions by Iraq into Kuwait again created tension; US-led air strikes restored calm.

# Kyrgyzstan
Republic of

*area* 198,500 sq km/76,641 sq mi
*capital* Bishkek (formerly Frunze)
*towns* Osh, Przhevalsk, Kyzyl-Kiya, Tormak
*physical* mountainous, an extension of the Tian Shan range
*head of state* Askar Akayev from 1990
*head of government* Tursunbek Chyngyshev from 1991
*political system* emergent democracy
*products* cereals, sugar, cotton, coal, oil, sheep, yaks, horses
*population* (1990) 4,400,000 (Kyrgyz 52%, Russian 22%, Uzbek 13%, Ukrainian 3%, German 2%)
*language* Kyrgyz, a Turkic language
*religion* Sunni Muslim
*chronology*
*1917–1924* Part of an independent Turkestan republic.
*1924* Became autonomous republic within USSR.
*1936* Became full union republic within USSR.
*1990* June: ethnic clashes resulted in state of emergency being imposed in Bishkek. Nov: Askar Akayev chosen as state president.

*1991* March: Kyrgyz voters endorsed maintenance of Union in USSR referendum. Aug: President Akayev condemned anti-Gorbachev attempted coup in Moscow; Kyrgyz Communist Party, which supported the coup, suspended. Oct: Akayev directly elected president. Dec: joined new Commonwealth of Independent States (CIS) and independence recognized by USA.
*1992* Jan: admitted into Conference on Security and Cooperation in Europe (CSCE); March: became a member of the United Nations. Dec: Supreme Soviet (parliament) renamed the Uluk Kenesh.

## Laos
Lao People's
Democratic
Republic
(*Saathiaranagroat
Prachhathippatay
Prachhachhon Lao*)

*area* 236,790 sq km/91,400 sq mi
*capital* Vientiane
*towns* Luang Prabang (the former royal capital), Pakse, Savannakhet
*physical* landlocked state with high mountains in E; Mekong River in W; jungle covers nearly 60% of land
*head of state* Nouhak Phoumsavan from 1992
*head of government* General Khamtay Siphandon from 1991
*political system* communism, one-party state
*exports* hydroelectric power from the Mekong is exported to Thailand, timber, teak, coffee, electricity
*currency* new kip
*population* (1990 est) 4,024,000 (Lao 48%, Thai 14%, Khmer 25%, Chinese 13%); growth rate 2.2% p.a.
*life expectancy* men 48, women 51 (1989)
*languages* Lao (official), French
*religions* Theravāda Buddhist 85%, animist beliefs among mountain dwellers

*literacy* 45% (1991)
*GNP* $500 million (1987); $180 per head (1988)
*chronology*
*1893–1945* Laos was a French protectorate.
*1945* Temporarily occupied by Japan.
*1946* Retaken by France.
*1950* Granted semi-autonomy in French Union.
*1954* Independence achieved from France.
*1960* Right-wing government seized power.
*1962* Coalition government established; civil war continued.
*1973* Vientiane cease-fire agreement. Withdrawal of US, Thai, and North Vietnamese forces.
*1975* Communist-dominated republic proclaimed with Prince Souphanouvong as head of state.
*1986* Phoumi Vongvichit became acting president.
*1988* Plans announced to withdraw 40% of Vietnamese forces stationed in the country.
*1989* First assembly elections since communist takeover.
*1991* Constitution approved. Kaysone Phomvihane elected president. General Khamtay Siphandon named as new premier.
*1992* Question of US prisoners of war retained in Laos since the end of Vietnam War unresolved. Nov: Phomvihane died; replaced by Nouhak Phoumsavan. Dec: new national assembly created, replacing supreme people's assembly, and general election held (effectively one-party).

# Latvia
Republic of

*area* 63,700 sq km/24,595 sq mi
*capital* Riga
*towns* Daugavpils, Liepāja, Jurmala, Jelgava, Ventspils
*physical* wooded lowland (highest point 312 m/1,024 ft), marshes, lakes; 472 km/293 mi of coastline; mild climate
*head of state* Anatolijs Gorbunov from 1988
*head of government* Ivars Godmanis from 1990
*political system* emergent democratic republic
*products* electronic and communications equipment, electric railway carriages, motorcycles, consumer durables, timber, paper and woollen goods, meat and dairy products
*currency* Latvian rouble
*population* (1990) 2,700,000 (Latvian 52%, Russian 34%, Byelorussian 5%, Ukrainian 3%)
*language* Latvian
*religions* mostly Lutheran Protestant, with a Roman Catholic minority
*chronology*
*1917* Soviets and Germans contested for control of Latvia.
*1918* Soviet forces overthrown by Germany. Latvia declared independence. Soviet rule restored after German withdrawal.

***1919*** Soviet rule overthrown by British naval and German forces May–Dec; democracy established.
***1934*** Coup replaced established government.
***1939*** German-Soviet secret agreement placed Latvia under Russian influence.
***1940*** Incorporated into USSR as constituent republic.
***1941–44*** Occupied by Germany.
***1944*** USSR regained control.
***1980*** Nationalist dissent began to grow.
***1988*** Latvian Popular Front established to campaign for independence. Prewar flag readopted; official status given to Latvian language.
***1989*** Popular Front swept local elections.
***1990*** Jan: Communist Party's monopoly of power abolished. March–April: Popular Front secured majority in elections. April: Latvian Communist Party split into pro-independence and pro-Moscow wings. May: unilateral declaration of independence from USSR, subject to transitional period for negotiation.
***1991*** Jan: Soviet troops briefly seized key installations in Riga. March: overwhelming vote for independence in referendum.
Aug: full independence declared at time of anti-Gorbachev coup; Communist Party outlawed. Sept: independence recognized by Soviet government and Western nations; United Nations (UN) membership granted; admitted into Conference on Security and Cooperation in Europe (CSCE).
***1992*** US reopened its embassy in Latvia. Russia began pullout of ex-Soviet troops, to be completed 1994. July: curbing of rights of non-citizens in Latvia prompted Russia to request minority protection by UN.

## Lebanon
Republic of
(*al-Jumhouria al-Lubnaniya*)

*area* 10,452 sq km/4,034 sq mi
*capital* Beirut
*towns* Tripoli, Tyre, Sidon
*physical* narrow coastal plain; Bekka valley N–S between Lebanon and Anti-Lebanon mountain ranges
*head of state* Elias Hrawi from 1989
*head of government* Rafik al-Hariri from 1992
*political system* emergent democratic republic
*exports* citrus and other fruit, vegetables; industrial products to Arab neighbours
*currency* Lebanese pound
*population* (1990 est) 3,340,000 (Lebanese 82%, Palestinian 9%, Armenian 5%); growth rate –0.1% p.a.
*life expectancy* men 65, women 70 (1989)
*languages* Arabic, French (both official), Armenian, English
*religions* Muslim 57% (Shiite 33%, Sunni 24%), Christian (Maronite and Orthodox) 40%, Druse 3%
*literacy* 75% (1989)
*GNP* $1.8 bn; $690 per head (1986)

## Lebanon

*chronology*

*1920–41* Administered under French mandate.
*1944* Independence achieved.
*1948–49* Lebanon joined first Arab war against Israel. Palestinian refugees settled in the south.
*1964* Palestine Liberation Organization (PLO) founded in Beirut.
*1971* PLO expelled from Jordan; established headquarters in Lebanon.
*1975* Outbreak of civil war between Christians and Muslims.
*1976* Cease-fire agreed; Syrian-dominated Arab deterrent force formed to keep the peace but considered by Christians as a occupying force.
*1978* Israel invaded S Lebanon in search of PLO fighters. International peacekeeping force established. Fighting broke out again.
*1979* Part of S Lebanon declared an 'independent free Lebanon'.
*1982* Bachir Gemayel became president but was assassinated before he could assume office; succeeded by his brother Amin Gemayel. Israel again invaded Lebanon. Palestinians withdrew from Beirut under supervision of international peacekeeping force.
*1983* Agreement reached for the withdrawal of Syrian and Israeli troops but abrogated under Syrian pressure.
*1984* Most of international peacekeeping force withdrawn. Muslim militia took control of W Beirut.
*1985* Lebanon in chaos; many foreigners taken hostage.
*1987* Syrian troops sent into Beirut.
*1989* Christian leader General Michel Aoun declared 'war of liberation' against Syrian occupation; Saudi Arabia and Arab League sponsored talks that resulted in new constitution recognizing Muslim majority; René Muhawad named president, assassinated after 17 days in office; Elias Hrawi named successor; Aoun occupied presidential palace, rejected constitution.
*1990* Release of Western hostages began. Aoun surrendered and legitimate government restored, with Umar Karami as prime minister.
*1991* Government extended control to the whole country. Treaty of cooperation with Syria signed. More Western hostages released.
*1992* Karami succeeded by Rashid al-Solh. Remaining Western hostages released. General election boycotted by many Christians; pro-Syrian administration reelected; Rafik al-Hariri prime minister.

# Lesotho
Kingdom of

*area* 30,355 sq km/11,717 sq mi
*capital* Maseru
*towns* Teyateyaneng, Mafeteng, Roma, Quthing
*physical* mountainous with plateaus, forming part of South Africa's chief watershed
*political system* emergent democracy
*head of state* King Letsie III from 1990
*head of government* Ntsu Mokhehle from 1993
*exports* wool, mohair, diamonds, cattle, wheat, vegetables
*currency* maluti
*population* (1990 est) 1,757,000; growth rate 2.7% p.a.
*life expectancy* men 59, women 62 (1989)
*languages* Sesotho, English (official), Zulu, Xhosa
*religions* Protestant 42%, Roman Catholic 38%
*literacy* 59% (1988)
*GNP* $408 million; $410 per head (1988)
*chronology*
*1868* Basutoland became a British protectorate.

*1966* Independence achieved from Britain, within the Commonwealth, as the Kingdom of Lesotho, with Moshoeshoe II as king and Chief Leabua Jonathan as prime minister.
*1970* State of emergency declared and constitution suspended.
*1973* Progovernment interim assembly established; Basotha National Party (BNP) won majority of seats.
*1975* Members of the ruling party attacked by guerrillas backed by South Africa.
*1985* Elections cancelled because no candidates opposed BNP.
*1986* South Africa imposed border blockade, forcing deportation of 60 African National Congress members. General Lekhanya ousted Chief Jonathan in coup. National assembly abolished. Highlands Water Project agreement signed with South Africa.
*1990* Moshoeshoe II dethroned by military council; replaced by his son Mohato as King Letsie III.
*1991* Lekhanya ousted in military coup led by Col Elias Tutsoane Ramaema. Political parties permitted to operate.
*1992* Ex-king Moshoeshoe returned from exile.
*1993* Free elections ended military rule; Ntsu Mokhehle of Basutoland Congress Party (BCP) became prime minister.

# Liberia
Republic of

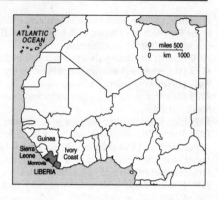

*area* 111,370 sq km/42,989 sq mi
*capital* Monrovia
*towns* Buchanan, Greenville
*physical* forested highlands; swampy tropical coast where six rivers enter the sea
*head of state and government* Amos Sawyer from 1990
*political system* emergent democratic republic
*exports* iron ore, rubber (Africa's largest producer), timber, diamonds, coffee, cocoa, palm oil
*currency* Liberian dollar
*population* (1990 est) 2,644,000 (95% indigenous); growth rate 3% p.a.
*life expectancy* men 53, women 56 (1989)
*languages* English (official), over 20 Niger-Congo languages
*religions* animist 65%, Muslim 20%, Christian 15%
*literacy* men 47%, women 23% (1985 est)
*GNP* $973 million; $410 per head (1987)
*chronology*
*1847* Founded as an independent republic.
*1944* William Tubman elected president.
*1971* Tubman died; succeeded by William Tolbert.

*1980* Tolbert assassinated in coup led by Samuel Doe, who suspended the constitution and ruled through a People's Redemption Council.
*1984* New constitution approved. National Democratic Party of Liberia (NDPL) founded by Doe.
*1985* NDPL won decisive victory in general election. Unsuccessful coup against Doe.
*1990* Rebels under former government minister Charles Taylor controlled nearly entire country by July. Doe killed during a bloody civil war between rival rebel factions. Amos Sawyer became interim head of government.
*1991* Amos Sawyer re-elected president. Rebel leader Charles Taylor agreed to work with Sawyer. Peace agreement failed but later revived; UN peacekeeping force drafted into republic.
*1992* Monrovia under siege by Taylor's rebel forces.

# Libya

**Libya**
Great Socialist
People's Libyan
Arab Jamahiriya
(*al-Jamahiriya
al-Arabiya al-Libya
al-Shabiya
al-Ishtirakiya
al-Uzma*)

*area* 1,759,540 sq km/679,182 sq mi
*capital* Tripoli
*towns* ports Benghazi, Misurata, Tobruk
*physical* flat to undulating plains with plateaus and depressions stretch S from the Mediterranean coast to an extremely dry desert interior
*environment* plan to pump water from below the Sahara to the coast risks rapid exhaustion of nonrenewable supply (Great Manmade River Project)
*head of state and government* Moamer al-Khaddhafi from 1969
*political system* one-party socialist state
*exports* oil, natural gas
*currency* Libyan dinar
*population* (1990 est) 4,280,000 (including 500,000 foreign workers); growth rate 3.1% p.a.
*life expectancy* men 64, women 69 (1989)
*language* Arabic
*religion* Sunni Muslim 97%
*literacy* 60% (1989)
*GNP* $20 bn; $5,410 per head (1988)

*chronology*
*1911* Conquered by Italy.
*1934* Colony named Libya.
*1942* Divided into three provinces: Fezzan (under French control); Cyrenaica, Tripolitania (under British control).
*1951* Achieved independence as the United Kingdom of Libya, under King Idris.
*1969* King deposed in a coup led by Col Moamer al-Khaddhafi. Revolution Command Council set up and the Arab Socialist Union (ASU) proclaimed the only legal party.
*1972* Proposed federation of Libya, Syria, and Egypt abandoned.
*1980* Proposed merger with Syria abandoned. Libyan troops began fighting in Chad.
*1981* Proposed merger with Chad abandoned.
*1986* US bombing of Khaddhafi's headquarters, following allegations of his complicity in terrorist activities.
*1988* Diplomatic relations with Chad restored.
*1989* USA accused Libya of building a chemical-weapons factory and shot down two Libyan planes; reconciliation with Egypt.
*1992* Khaddhafi under international pressure to extradite suspected Lockerbie and UTA (Union de Transports Aeriens) bombers for trial outside Libya; sanctions imposed.

# Liechtenstein

**Liechtenstein**
Principality of
(*Fürstentum Liechtenstein*)

*area* 160 sq km/62 sq mi
*capital* Vaduz
*towns* Balzers, Schaan, Ruggell
*physical* landlocked Alpine; includes part of Rhine Valley in W
*head of state* Prince Hans Adam II from 1989
*head of government* Hans Brunhart from 1978
*political system* constitutional monarchy
*exports* microchips, dental products, small machinery, processed foods, postage stamps
*currency* Swiss franc
*population* (1990 est) 30,000 (33% foreign); growth rate 1.4% p.a.
*life expectancy* men 78, women 83 (1989)
*language* German (official); an Alemannic dialect is also spoken
*religions* Roman Catholic 87%, Protestant 8%
*literacy* 100% (1989)
*GNP* $450 million (1986)
*chronology*
*1342* Became a sovereign state.

*1434* Present boundaries established.
*1719* Former counties of Schellenberg and Vaduz constituted as the Principality of Liechtenstein.
*1921* Adopted Swiss currency.
*1923* United with Switzerland in a customs union.
*1938* Prince Franz Josef II came to power.
*1984* Prince Franz Joseph II handed over power to Crown Prince Hans Adam. Vote extended to women in national elections.
*1989* Prince Franz Joseph II died; Hans Adam II succeeded him.
*1990* Became a member of the United Nations (UN).
*1991* Became seventh member of European Free Trade Association.

# Lithuania
Republic of

*area* 65,200 sq km/25,174 sq mi
*capital* Vilnius
*towns* Kaunas, Klaipeda, Siauliai, Panevezys
*physical* central lowlands with gentle hills in W and higher terrain in SE; 25% forested; some 3,000 small lakes, marshes, and complex sandy coastline
*head of state* Algirdas Brazauskas from 1993
*head of government* Bronislovas Lubys from 1992
*political system* emergent democracy
*products* heavy engineering, electrical goods, shipbuilding, cement, food processing, bacon, dairy products, cereals, potatoes
*currency* Lithuanian rouble
*population* (1990) 3,700,000 (Lithuanian 80%, Russian 9%, Polish 7%, Byelorussian 2%)
*language* Lithuanian
*religion* predominantly Roman Catholic
*chronology*
*1918* Independence declared following withdrawal of German occupying troops at end of World War I; USSR attempted to regain power.

*1919* Soviet forces overthrown by Germans, Poles, and nationalist Lithuanians; democratic republic established.

*1920–39* Province and city of Vilnius occupied by Poles.

*1926* Coup overthrew established government; Antanas Smetona became president.

*1939* Secret German-Soviet agreement brought most of Lithuania under Soviet influence.

*1940* Incorporated into USSR as constituent republic.

*1941* Lithuania revolted against USSR and established own government. During World War II Germany again occupied the country.

*1944* USSR resumed rule.

*1944–52* Lithuanian guerrillas fought USSR.

*1972* Demonstrations against Soviet government.

*1980* Growth in nationalist dissent, influenced by Polish example.

*1988* Popular front formed, the Sajudis, to campaign for increased autonomy.

*1989* Lithuanian declared the state language; flag of independent interwar republic readopted. Communist Party (CP) split into pro-Moscow and nationalist wings. Communist local monopoly of power abolished.

*1990* Feb: nationalist Sajudis won elections. March: Vytautas Landsbergis became president; unilateral declaration of independence resulted in temporary Soviet blockade.

*1991* Jan: Albertas Shiminas became prime minister. Soviet paratroopers briefly occupied key buildings in Vilnius. Sept: independence recognized by Soviet government and Western nations; Gediminas Vagnorius elected prime minister; CP outlawed; admitted into United Nations and Conference on Security and Cooperation in Europe (CSCE).

*1992* July: Aleksandras Abisala became prime minister. Nov: Democratic Labour Party, led by Algirdas Brazauskas, won majority vote. Dec: Bronislovas Lubys appointed prime minister.

*1993* Brazauskas elected president.

## Luxembourg
Grand Duchy of
(*Grand-Duché de Luxembourg*)

*area* 2,586 sq km/998 sq mi
*capital* Luxembourg
*towns* Esch-sur-Alzette, Dudelange
*physical* on the river Moselle; part of the Ardennes (Oesling) forest in N
*head of state* Grand Duke Jean from 1964
*head of government* Jacques Santer from 1984
*political system* liberal democracy
*exports* pharmaceuticals, synthetic textiles, steel
*currency* Luxembourg franc
*population* (1990 est) 369,000; growth rate 0% p.a.
*life expectancy* men 71, women 78 (1989)
*languages* French (official), local Letzeburgesch, German
*religion* Roman Catholic 97%
*literacy* 100% (1989)
*GNP* $4.9 bn; $13,380 per head (1988)
*chronology*
*1354* Became a duchy.

*1482* Under Habsburg control.
*1797* Ceded, with Belgium, to France.
*1815* Treaty of Vienna created Luxembourg a grand duchy, ruled by the king of the Netherlands.
*1830* With Belgium, revolted against Dutch rule.
*1890* Link with Netherlands ended with accession of Grand Duke Adolphe of Nassau-Weilburg.
*1948* With Belgium and the Netherlands, formed the Benelux customs union.
*1960* Benelux became fully effective economic union.
*1961* Prince Jean became acting head of state on behalf of his mother, Grand Duchess Charlotte.
*1964* Grand Duchess Charlotte abdicated; Prince Jean became grand duke.
*1974* Dominance of Christian Social Party challenged by Socialists.
*1979* Christian Social Party regained pre-eminence.
*1991* Pact agreeing European free-trade area signed in Luxembourg.
*1992* Voted in favour of ratification of Maastricht Treaty on European union.

# Madagascar

**Madagascar**
Democratic Republic of
(*Repoblika
Demokratika n'i
Madagaskar*)

*area* 587,041 sq km/226,598 sq mi
*capital* Antananarivo
*towns* Toamasina, Antseranana, Fianarantsoa, Toliary
*physical* temperate central highlands; humid valleys and tropical coastal plains; arid in S
*environment* according to 1990 UN figures, 93% of the forest area has been destroyed and about 100,000 species have been made extinct
*head of state* Albert Zafy from 1993
*head of government* Guy Razanamasy from 1991
*political system* emergent democratic republic
*exports* coffee, cloves, vanilla, sugar, chromite, shrimps
*currency* Malagasy franc
*population* (1990 est) 11,802,000, mostly of Malayo-Indonesian origin; growth rate 3.2% p.a.
*life expectancy* men 50, women 53 (1989)
*languages* Malagasy (official), French, English
*religions* animist 50%, Christian 40%, Muslim 10%

*literacy* 53% (1988)
*GNP* $2.1 bn (1987); $280 per head (1988)
*chronology*
*1885* Became a French protectorate.
*1896* Became a French colony.
*1960* Independence achieved from France, with Philibert Tsiranana as president.
*1972* Army took control of the government.
*1975* Martial law imposed under a national military directorate. New Marxist constitution proclaimed the Democratic Republic of Madagascar, with Didier Ratsiraka as president.
*1976* Front-Line Revolutionary Organization (AREMA) formed.
*1977* National Front for the Defence of the Malagasy Socialist Revolution (FNDR) became the sole legal political organization.
*1980* Ratsiraka abandoned Marxist experiment.
*1983* Ratsiraka re-elected, despite strong opposition from radical socialist National Movement for the Independence of Madagascar (MONIMA) under Monja Jaona.
*1989* Ratsiraka re-elected for third term after restricting opposition parties.
*1990* Political opposition legalized; 36 new parties created.
*1991* Antigovernment demonstrations; opposition to Ratsiraka led to general strike. Nov: Ratsiraka formed new unity government.
*1992* Constitutional reform approved. Oct: first multiparty elections won by Democrat coalition.
*1993* Albert Zafy, leader of coalition, elected president.

# Malawi
Republic of
(*Malaŵi*)

*area* 118,000 sq km/45,560 sq mi
*capital* Lilongwe
*towns* Blantyre, Mzuzu, Zomba
*physical* landlocked narrow plateau with rolling plains; mountainous W of Lake Malawi
*head of state and government* Hastings Kamusu Banda from 1966 for life
*political system* one-party republic
*exports* tea, tobacco, cotton, peanuts, sugar
*currency* kwacha
*population* (1990 est) 9,080,000 (nearly 1 million refugees from Mozambique); growth rate 3.3% p.a.
*life expectancy* men 46, women 50 (1989)
*languages* English, Chichewa (both official)
*religions* Christian 75%, Muslim 20%
*literacy* 25% (1989)
*GNP* $1.2 bn (1987); $160 per head (1988)

*chronology*
*1891* Became the British protectorate Nyasaland.
*1964* Independence achieved from Britain, within the Commonwealth, as Malawi.
*1966* Became a one-party republic, with Hastings Banda as president.
*1971* Banda was made president for life.
*1977* Banda released some political detainees and allowed greater freedom of the press.
*1986–89* Influx of nearly a million refugees from Mozambique.
*1992* Calls for multiparty politics. Countrywide industrial riots caused many fatalities. Western aid suspended over human-rights violations. Referendum on constitutional reform promised.
*1993* Commission appointed to supervise preparations for the referendum.

## Malaysia

*area* 329,759 sq km/127,287 sq mi
*capital* Kuala Lumpur
*towns* Johor Baharu, Ipoh, Georgetown (Penang), Kuching in Sarawak, Kota Kinabalu in Sabah
*physical* comprises Peninsular Malaysia (the nine Malay states—Johore, Kedah, Kelantan, Negri Sembilan, Pahang, Perak, Perlis, Selangor, Trengganu—plus Malacca and Penang); and E Malaysia (Sabah and Sarawak); 75% tropical jungle; central mountain range; swamps in E
*head of state* Rajah Azlan Muhibuddin Shah (sultan of Perak) from 1989
*head of government* Mahathir bin Mohamad from 1981
*political system* liberal democracy
*exports* pineapples, palm oil, rubber, timber, petroleum (Sarawak), bauxite
*currency* ringgit
*population* (1990 est) 17,053,000 (Malaysian 47%, Chinese 32%,

Indian 8%, others 13%); growth rate 2% p.a.
*life expectancy* men 65, women 70 (1989)
*languages* Malay (official), English, Chinese, Indian, and local languages
*religions* Muslim (official), Buddhist, Hindu, local beliefs
*literacy* 80% (1989)
*GNP* $34.3 bn; $1,870 per head (1988)
*chronology*
*1786* Britain established control.
*1826* Became a British colony.
*1963* Federation of Malaysia formed, including Malaya, Singapore, Sabah (N Borneo), and Sarawak (NW Borneo).
*1965* Secession of Singapore from federation.
*1969* Anti-Chinese riots in Kuala Lumpur.
*1971* Launch of *bumiputra*, ethnic-Malay-oriented economic policy.
*1981* Election of Dr Mahathir bin Mohamad as prime minister.
*1982* Mahathir bin Mohamad re-elected.
*1986* Mahathir bin Mohamad re-elected.
*1987* Arrest of over 100 opposition activists, including Democratic Action Party (DAP) leader, as Malay-Chinese relations deteriorated.
*1988* Split in ruling New United Malays' National Organization (UMNO) party over Mahathir's leadership style; new UMNO formed.
*1989* Semangat '46 set up by former members of UMNO including ex-premier Tunku Abdul Rahman.
*1990* Mahathir bin Mohamad re-elected.
*1991* New economic growth programme launched.

# Maldives
Republic of
(*Divehi Jumhuriya*)

*area* 298 sq km/115 sq mi
*capital* Malé
*towns* Seenu
*physical* comprises 1,196 coral islands, grouped into 12 clusters of atolls, largely flat, none bigger than 13 sq km/5 sq mi, average elevation 1.8 m/6 ft; 203 are inhabited
*environment* the threat of rising sea level has been heightened by the frequency of flooding in recent years
*head of state and government* Maumoon Abdul Gayoom from 1978
*political system* authoritarian nationalism
*exports* coconuts, copra, bonito (fish related to tuna), garments
*currency* Rufiya
*population* (1990 est) 219,000; growth rate 3.7% p.a.
*life expectancy* men 60, women 63 (1989)
*languages* Divehi (Sinhalese dialect), English
*religion* Sunni Muslim
*literacy* 36% (1989)
*GNP* $69 million (1987); $410 per head (1988)

*chronology*
*1887* Became a British protectorate.
*1953* Long a sultanate, the Maldive Islands became a republic within the Commonwealth.
*1954* Sultan restored.
*1965* Achieved full independence outside the Commonwealth.
*1968* Sultan deposed; republic reinstated with Ibrahim Nasir as president.
*1978* Nasir retired; replaced by Maumoon Abdul Gayoom.
*1982* Rejoined the Commonwealth.
*1983* Gayoom re-elected.
*1985* Became a founder member of South Asian Association for Regional Cooperation (SAARC).
*1988* Gayoom re-elected. Coup attempt by mercenaries, thought to have the backing of former president Nasir, was foiled by Indian paratroops.

## Mali
Republic of
(*République du Mali*)

*area* 1,240,142 sq km/478,695 sq mi
*capital* Bamako
*towns* Mopti, Kayes, Ségou, Timbuktu
*physical* landlocked state with river Niger and savanna in S; part of the Sahara in N; hills in NE; Senegal River and its branches irrigate the SW
*environment* a rising population coupled with recent droughts has affected marginal agriculture. Once in surplus, Mali has had to import grain every year since 1965
*head of state and government* Alpha Oumar Konare from 1992
*political system* emergent democratic republic
*exports* cotton, peanuts, livestock, fish
*currency* franc CFA
*population* (1990 est) 9,182,000; growth rate 2.9% p.a.
*life expectancy* men 44, women 47 (1989)
*languages* French (official), Bambara
*religion* Sunni Muslim 90%, animist 9%, Christian 1%
*literacy* 10% (1989)
*GNP* $1.6 bn (1987); $230 per head (1988)
*chronology*
*1895* Came under French rule.

*1959* With Senegal, formed the Federation of Mali.
*1960* Became the independent Republic of Mali, with Modibo Keita as president.
*1968* Keita replaced in an army coup by Moussa Traoré.
*1974* New constitution made Mali a one-party state.
*1976* New national party, the Malian People's Democratic Union, announced.
*1983* Agreement between Mali and Guinea for eventual political and economic integration signed.
*1985* Conflict with Burkina Faso lasted five days; mediated by International Court of Justice.
*1991* Demonstrations against one-party rule. Moussa Traoré ousted in a coup led by Lt-Col Amadou Toumani Toure. New multiparty constitution agreed, subject to referendum.
*1992* Referendum endorsed new democratic constitution. Alliance for Democracy in Mali (ADEMA) won multiparty elections; Alpha Oumar Konare elected president.

# Malta

**Malta**
Republic of
(*Repubblika Ta'Malta*)

*area* 320 sq km/124 sq mi
*capital* Valletta
*towns* Rabat, Marsaxlokk
*physical* includes islands of Gozo 67 sq km/26 sq mi and Comino 2.5 sq km/1 sq mi
*head of state* Vincent Tabone from 1989
*head of government* Edward Fenech Adami from 1987
*political system* liberal democracy
*exports* vegetables, knitwear, handmade lace, plastics, electronic equipment
*currency* Maltese lira
*population* (1990 est) 373,000; growth rate 0.7% p.a.
*life expectancy* men 72, women 77 (1987)
*languages* Maltese, English
*religion* Roman Catholic 98%
*literacy* 90% (1988)
*GNP* $1.6 bn; $4,750 per head (1988)
*chronology*
*1814* Annexed to Britain by the Treaty of Paris.

*1947* Achieved self-government.
*1955* Dom Mintoff of the Malta Labour Party (MLP) became prime minister.
*1956* Referendum approved MLP's proposal for integration with the UK. Proposal opposed by the Nationalist Party.
*1958* MLP rejected the British integration proposal.
*1962* Nationalists elected, with Borg Olivier as prime minister.
*1964* Independence achieved from Britain, within the Commonwealth. Ten-year defence and economic-aid treaty with UK signed.
*1971* Mintoff re-elected. 1964 treaty declared invalid and negotiations began for leasing the NATO base in Malta.
*1972* Seven-year NATO agreement signed.
*1974* Became a republic.
*1979* British military base closed.
*1984* Mintoff retired and was replaced by Mifsud Bonnici as prime minister and MLP leader.
*1987* Edward Fenech Adami (Nationalist) elected prime minister.
*1989* Vincent Tabone elected president. USA–USSR summit held offshore.
*1990* Formal application made for EC membership.
*1992* Nationalist Party returned to power in general election.

# Mauritania

**Mauritania**
Islamic Republic of
(*République Islamique de Mauritanie*)

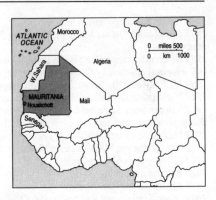

*area* 1,030,700 sq km/397,850 sq mi
*capital* Nouakchott
*towns* Nouadhibou, Kaédi, Zouérate
*physical* valley of river Senegal in S; remainder arid and flat
*head of state and government* Maaouia Ould Sid Ahmed Taya from 1984
*political system* emergent democratic republic
*exports* iron ore, fish, gypsum
*currency* ouguiya
*population* (1990 est) 2,038,000 (Arab-Berber 30%, black African 30%, Haratine 30%—descendants of black slaves, who remained slaves until 1980); growth rate 3% p.a.
*life expectancy* men 43, women 48 (1989)
*languages* French (official), Hasaniya Arabic, black African languages
*religion* Sunni Muslim 99%
*literacy* 17% (1987)
*GNP* $843 million; $480 per head (1988)
*chronology*
*1903* Became a French protectorate.
*1960* Independence achieved from France, with Moktar Ould Daddah as president.

*1975* Western Sahara ceded by Spain. Mauritania occupied the southern area and Morocco the north. Polisario Front formed in Sahara to resist the occupation by Mauritania and Morocco.
*1978* Daddah deposed in bloodless coup; replaced by Mohamed Khouna Ould Haidalla. Peace agreed with Polisario Front.
*1981* Diplomatic relations with Morocco broken.
*1984* Haidalla overthrown by Maaouia Ould Sid Ahmed Taya. Polisario regime formally recognized.
*1985* Relations with Morocco restored.
*1989* Violent clashes between Mauritanians and Senegalese. Arab-dominated government expelled thousands of Africans into N Senegal; governments had earlier agreed to repatriate each other's citizens (about 250,000).
*1991* Amnesty for political prisoners. Multiparty elections promised. Calls for resignation of President Taya.
*1992* First multiparty elections won by ruling Democratic and Social Republican Party (PRDS). Diplomatic relations with Senegal resumed.

## Mauritius
Republic of

*area* 1,865 sq km/720 sq mi; the island of Rodrigues is part of Mauritius; there are several small island dependencies
*capital* Port Louis
*towns* Beau Bassin-Rose Hill, Curepipe, Quatre Bornes
*physical* mountainous, volcanic island surrounded by coral reefs
*interim head of state* Veerasamy Ringadoo from 1992
*head of government* Aneerood Jugnauth from 1982
*political system* liberal democratic republic
*exports* sugar, knitted goods, tea
*currency* Mauritius rupee
*population* (1990 est) 1,141,900, 68% of Indian origin; growth rate 1.5% p.a.
*life expectancy* men 64, women 71 (1989)
*languages* English (official), French, creole, Indian languages
*religions* Hindu 51%, Christian 30%, Muslim 17%
*literacy* 94% (1989)
*GNP* $1.4 bn (1987); $1,810 per head (1988)
*chronology*
*1814* Annexed to Britain by the Treaty of Paris.
*1968* Independence achieved from Britain within the Commonwealth, with Seewoosagur Ramgoolam as prime minister.

*1982* Aneerood Jugnauth became prime minister.
*1983* Jugnauth formed a new party, the Mauritius Socialist Movement. Ramgoolam appointed governor general. Jugnauth formed a new coalition government.
*1985* Ramgoolam died; succeeded by Veersamy Ringadoo.
*1987* Jugnauth's coalition re-elected.
*1990* Attempt to create a republic failed.
*1991* Jugnauth's ruling Mauritius Socialist Movement–Mauritius Militant Movement–Rodriguais People's Organization coalition won general election; pledge to secure republican status by 1992.
*1992* Mauritius became a republic while remaining a member of the Commonwealth. Ringadoo became interim president.

# Mexico
United States of
(*Estados Unidos Mexicanos*)

*area* 1,958,201 sq km/756,198 sq mi
*capital* Mexico City
*towns* Guadalajara, Monterrey, Veracruz
*physical* partly arid central highlands; Sierra Madre mountain ranges E and W; tropical coastal plains
*environment* during the 1980s, smog levels in Mexico City exceeded World Health Organization standards on more than 300 days of the year. Air is polluted by 130,000 factories and 2.5 million vehicles
*head of state and government* Carlos Salinas de Gortari from 1988
*political system* federal democratic republic
*exports* silver, gold, lead, uranium, oil, natural gas, handicrafts, fish, shellfish, fruits and vegetables, cotton, machinery
*currency* peso
*population* (1990 est) 88,335,000 (mixed descent 60%, Indian 30%, Spanish descent 10%); 50% under 20 years of age; growth rate 2.6% p.a.
*life expectancy* men 67, women 73
*languages* Spanish (official) 92%, Nahuatl, Maya, Mixtec
*religion* Roman Catholic 97%

*literacy* men 92%, women 88% (1989)
*GNP* $126 bn (1987); $2,082 per head
*chronology*
*1821* Independence achieved from Spain.
*1846–48* Mexico at war with USA; loss of territory.
*1848* Maya Indian revolt suppressed.
*1864–67* Maximilian of Austria was emperor of Mexico.
*1917* New constitution introduced, designed to establish permanent democracy.
*1983–84* Financial crisis.
*1985* Institutional Revolutionary Party (PRI) returned to power. Earthquake in Mexico City.
*1986* International Monetary Fund loan agreement signed to keep the country solvent until at least 1988.
*1988* PRI candidate Carlos Salinas de Gortari elected president. Debt reduction accords negotiated with USA.
*1991* PRI won general election. President Salinas promised constitutional reforms.
*1992* Public outrage following Guadalajara gas-explosion disaster in which 194 people died and 1,400 were injured.

# Moldova
Republic of

*area* 33,700 sq km/13,012 sq mi
*capital* Chișinău (Kishinev)
*towns* Tiraspol, Beltsy, Bendery
*physical* hilly land lying largely between the rivers Prut and Dniester; northern Moldova comprises the level plain of the Beltsy Steppe and uplands; the climate is warm and moderately continental
*head of state* Mircea Snegur from 1989
*head of government* Valerin Murovsky from 1992
*political system* emergent democracy
*products* wine, tobacco, canned goods
*population* (1990) 4,400,000 (Moldavian 64%, Ukrainian 14%, Russian 13%, Gagauzi 4%, Bulgarian 2%)
*language* Moldavian, allied to Romanian
*religion* Russian Orthodox
*chronology*
*1940* Bessarabia in the E became part of the Soviet Union whereas the W part remained in Romania.
*1941* Bessarabia taken over by Romania–Germany.
*1944* Red army reconquered Bessarabia.
*1946–47* Widespread famine.

***1988*** A popular front, the Democratic Movement for Perestroika, campaigned for accelerated political reform.
***1989*** Jan–Feb: nationalist demonstrations in Chisinau. May: Moldavian Popular Front established. July: Mircea Snegur became head of state. Aug: Moldavian language granted official status triggering clashes between ethnic Russians and Moldavians. Nov: Gagauz-Khalky People's Movement formed to campaign for Gagauz autonomy.
***1990*** Feb: Popular Front polled strongly in supreme soviet elections. June: economic and political sovereignty declared; renamed Republic of Moldova. Oct: Gagauzi held unauthorized elections to independent parliament; state of emergency declared after inter-ethnic clashes. Trans-Dniester region declared its sovereignty. Nov: state of emergency declared in Trans-Dniester region after inter-ethnic killings.
***1991*** March: Moldova boycotted the USSR's constitutional referendum. Aug: independence declared after abortive anti-Gorbachev coup; Communist Party outlawed. Dec: Moldova joined new Commonwealth of Independent States (CIS).
***1992*** Jan: admitted into the Conference on Security and Cooperation in Europe (CSCE). Possible union with Romania discussed. March: state of emergency imposed; admitted into United Nations; diplomatic recognition granted by USA. May: further meeting on unification with Romania; Trans-Dniester region fighting intensified. July: Andrei Sangheli became premier; Moldova agreed to outside peacekeeping force. Aug: talks between Moldova and Russia began; Russian peacekeeping force reportedly deployed in Trans-Dniester region.

## Monaco
Principality of

***area*** 1.95 sq km/0.75 sq mi
***capital*** Monaco-Ville
***towns*** Monte Carlo, La Condamine, Fontvieille
***physical*** steep and rugged; surrounded landwards by French territory; being expanded by filling in the sea
***features*** aquarium and oceanographic centre; Monte Carlo film festival, motor races and casinos; world's second smallest state
***head of state*** Prince Rainier III from 1949
***head of government*** Jean Ausseil from 1986
***political system*** constitutional monarchy under French protectorate
***exports*** some light industry; economy dependent on tourism and gambling
***currency*** French franc
***population*** (1989) 29,000; growth rate –0.5% p.a.
***languages*** French (official), English, Italian
***religion*** Roman Catholic 95%
***literacy*** 99% (1985)
***chronology***
*1861* Became an independent state under French protection.

***1918*** France given a veto over succession to the throne.
***1949*** Prince Rainier III ascended the throne.
***1956*** Prince Rainier married US actress Grace Kelly.
***1958*** Birth of male heir, Prince Albert.
***1959*** Constitution of 1911 suspended.
***1962*** New constitution adopted.

# Mongolia

**Mongolia**
State of
(*Outer Mongolia* until 1924;
*People's Republic of Mongolia* until 1991)

*area* 1,565,000 sq km/604,480 sq mi
*capital* Ulaanbaatar
*towns* Darhan, Choybalsan
*physical* high plateau with desert and steppe (grasslands)
*head of state* Punsalmaagiyn Ochirbat from 1990
*head of government* Puntsagiyn Jasray from 1992
*political system* emergent democracy
*exports* meat and hides, minerals, wool, livestock, grain, cement, timber
*currency* tugrik
*population* (1990 est) 2,185,000; growth rate 2.8% p.a.
*life expectancy* men 63, women 67 (1989)
*languages* Khalkha Mongolian (official), Chinese, Russian, and Turkic languages
*religion* officially none (Tibetan Buddhist Lamaism suppressed 1930s)
*literacy* 89% (1985)
*GNP* $3.6 bn; $1,820 per head (1986)
*chronology*
*1911* Outer Mongolia gained autonomy from China.

***1915*** Chinese sovereignty reasserted.
***1921*** Chinese rule overthrown with Soviet help.
***1924*** People's Republic proclaimed.
***1946*** China recognized Mongolia's independence.
***1966*** 20-year friendship, cooperation, and mutual-assistance pact signed with USSR. Relations with China deteriorated.
***1984*** Yumjaagiyn Tsedenbal, effective leader, deposed and replaced by Jambyn Batmonh.
***1987*** Soviet troops reduced; Mongolia's external contacts broadened.
***1989*** Further Soviet troop reductions.
***1990*** Democratization campaign launched by Mongolian Democratic Union. Punsalmaagiyn Ochirbat's Mongolian People's Revolutionary Party (MPRP) elected in free multiparty elections. Mongolian script readopted.
***1991*** Massive privatization programme launched as part of move towards a market economy. The word 'Republic' dropped from country's name. GDP declined by 10%.
***1992*** Jan: New constitution introduced. Economic situation worsened; GDP again declined by 10%. Prime minister Dashiyn Byambasuren's resignation refused. July: Puntsagiyn Jasray appointed new prime minister.

## Morocco
Kingdom of
(*al-Mamlaka al-Maghrebia*)

*area* 458,730 sq km/177,070 sq mi (excluding Western Sahara)
*capital* Rabat
*towns* Marrakesh, Fez, Meknès, Casablanca, Tangier, Agadir
*physical* mountain ranges NE–SW; fertile coastal plains in W
*head of state* Hassan II from 1961
*head of government* Mohamed Lamrani from 1992
*political system* constitutional monarchy
*exports* dates, figs, cork, wood pulp, canned fish, phosphates
*currency* dirham (DH)
*population* (1990 est) 26,249,000; growth rate 2.5% p.a.
*life expectancy* men 62, women 65 (1989)
*languages* Arabic (official) 75%, Berber 25%, French, Spanish
*religion* Sunni Muslim 99%
*literacy* men 45%, women 22% (1985 est)
*GNP* $18.7 bn; $750 per head (1988)
*chronology*
*1912* Morocco divided into French and Spanish protectorates.
*1956* Independence achieved as the Sultanate of Morocco.
*1957* Sultan restyled king of Morocco.
*1961* Hassan II came to the throne.

***1969*** Former Spanish province of Ifni returned to Morocco.
***1972*** Major revision of the constitution.
***1975*** Western Sahara ceded by Spain to Morocco and Mauritania.
***1976*** Guerrilla war in Western Sahara with the Polisario Front. Sahrawi Arab Democratic Republic (SADR) established in Algiers. Diplomatic relations between Morocco and Algeria broken.
***1979*** Mauritania signed a peace treaty with Polisario.
***1983*** Peace formula for Western Sahara proposed by the Organization of African Unity (OAU); Morocco agreed but refused to deal directly with Polisario.
***1984*** Hassan signed an agreement for cooperation and mutual defence with Libya.
***1987*** Cease-fire agreed with Polisario, but fighting continued.
***1988*** Diplomatic relations with Algeria restored.
***1989*** Diplomatic relations with Syria restored.
***1992*** Mohamed Lamrani appointed prime minister; new constitution approved in national referendum.

## Mozambique
People's Republic of
(*República Popular de Moçambique*)

*area* 799,380 sq km/308,561 sq mi
*capital* (and chief port) Maputo
*towns* Beira, Nampula
*physical* mostly flat tropical lowland; mountains in W
*head of state and government* Joaquim Alberto Chissano from 1986
*political system* emergent democratic republic
*exports* prawns, cashews, sugar, cotton, tea, petroleum products, copra
*currency* metical (replaced escudo 1980)
*population* (1990 est) 14,718,000 (mainly indigenous Bantu peoples; Portuguese 50,000); growth rate 2.8% p.a.; nearly 1 million refugees in Malawi
*life expectancy* men 45, women 48 (1989)
*languages* Portuguese (official), 16 African languages
*religion* animist 60%, Roman Catholic 18%, Muslim 16%
*literacy* men 55%, women 22% (1985 est)
*GDP* $4.7 bn; $319 per head (1987)
*chronology*
*1505* Mozambique became a Portuguese colony.

*1962* Frelimo (liberation front) established.
*1975* Independence achieved from Portugal as a socialist republic, with Samora Machel as president and Frelimo as the sole legal party.
*1977* Renamo resistance group formed.
*1983* Re-establishment of good relations with Western powers.
*1984* Nkomati accord of nonaggression signed with South Africa.
*1986* Machel killed in air crash; succeeded by Joaquim Chissano.
*1988* Tanzania announced withdrawal of its troops. South Africa provided training for Mozambican forces.
*1989* Frelimo offered to abandon Marxist-Leninism; Chissano re-elected. Renamo continued attacks on government facilities and civilians.
*1990* One-party rule officially ended. Partial cease-fire agreed.
*1991* Peace talks resumed in Rome, delaying democratic process. Attempted antigovernment coup thwarted.
*1992* Aug: peace accord agreed upon, but fighting continued.
Oct: peace accord signed, but awaited ratification by government.

# Myanmar
Union of
(*Thammada Myanmar Naingngandaw*)
(formerly **Burma**, until 1989)

*area* 676,577 sq km/261,228 sq mi
*capital* (and chief port) Yangon (formerly Rangoon)
*towns* Mandalay, Moulmein, Pegu
*physical* over half is rainforest; rivers Irrawaddy and Chindwin in central lowlands ringed by mountains in N, W, and E
*environment* landslides and flooding during the rainy season (June–Sept) are becoming more frequent as a result of deforestation
*head of state and government* Than Shwe from 1992
*political system* military republic
*exports* rice, rubber, jute, teak, jade, rubies, sapphires
*currency* kyat
*population* (1990 est) 41,279,000; growth rate 1.9% p.a. (includes Shan, Karen, Raljome, Chinese, and Indian minorities)
*life expectancy* men 53, women 56 (1989)
*language* Burmese
*religions* Hinayana Buddhist 85%, animist, Christian
*literacy* 66% (1989)
*GNP* $9.3 bn (1988); $210 per head (1989)

*chronology*
*1886* United as province of British India.
*1937* Became crown colony in the British Commonwealth.
*1942–45* Occupied by Japan.
*1948* Independence achieved from Britain. Left the Commonwealth.
*1962* General Ne Win assumed power in army coup.
*1973–74* Adopted presidential-style 'civilian' constitution.
*1975* Opposition National Democratic Front formed.
*1986* Several thousand supporters of opposition leader Suu Kyi arrested.
*1988* Government resigned after violent demonstrations. General Saw Maung seized power in military coup Sept; over 1,000 killed.
*1989* Martial law declared; thousands arrested including advocates of democracy and human rights. Country renamed Myanmar and capital Yangon.
*1990* Breakaway opposition group formed 'parallel government' on rebel-held territory.
*1991* Martial law and human-rights abuses continued. Military offensives continued. Opposition leader, Aung San Suu Kyi, received Nobel Prize for Peace.
*1992* Jan–April: Pogrom against Muslim community in Arakan province, W Myanmar, carried out with army backing. April: Saw Maung replaced by Than Shwe. Several political prisoners liberated. Sept: martial law lifted, but restrictions on political freedom remained.
*1993* Constitutional convention held in Yangon to discuss adoption of proposed new constitution.

# Namibia

Republic of (formerly *South West Africa*)

*area* 824,300 sq km/318,262 sq mi
*capital* Windhoek
*towns* Swakopmund, Rehoboth, Rundu
*physical* mainly desert
*head of state* Sam Nujoma from 1990
*head of government* Hage Geingob from 1990
*political system* democratic republic
*exports* diamonds, uranium, copper, lead, zinc
*currency* South African rand
*population* (1990 est) 1,372,000 (black African 85%, European 6%)
*life expectancy* blacks 40, whites 69
*languages* Afrikaans (spoken by 60% of white population), German, English (all official), several indigenous languages
*religion* 51% Lutheran, 19% Roman Catholic, 6% Dutch Reformed Church, 6% Anglican
*literacy* whites 100%, nonwhites 16%
*GNP* $1.6 bn; $1,300 per head (1988)

*chronology*
*1884* German and British colonies established.
*1915* German colony seized by South Africa.
*1920* Administered by South Africa, under League of Nations mandate, as British South Africa.
*1946* Full incorporation in South Africa refused by United Nations.
*1958* South-West Africa People's Organization (SWAPO) set up to seek racial equality and full independence.
*1966* South Africa's apartheid laws extended to the country.
*1968* Redesignated Namibia by UN.
*1978* UN Security Council Resolution 435 for the granting of full sovereignty accepted by South Africa and then rescinded.
*1988* Peace talks between South Africa, Angola, and Cuba led to agreement on full independence for Namibia.
*1989* Unexpected incursion by SWAPO guerrillas from Angola into Namibia threatened agreed independence. Transitional constitution created by elected representatives; SWAPO dominant party.
*1990* Liberal multiparty 'independence' constitution adopted; independence achieved. Sam Nujoma elected president.
*1991* Agreement on joint administration of disputed port of Walvis Bay reached with South Africa, pending final settlement of dispute.
*1992* Agreement on establishment of Walvis Bay Joint Administrative Body.

## Nauru
Republic of
(*Naoero*)

*area* 21 sq km/8 sq mi
*capital* Yaren District
*physical* tropical island country in SW Pacific; plateau encircled by coral cliffs and sandy beaches
*head of state and government* Bernard Dowiyogo from 1989
*political system* liberal democracy
*exports* phosphates
*currency* Australian dollar
*population* (1990 est) 8,100 (mainly Polynesian; Chinese 8%, European 8%); growth rate 1.7% p.a.
*languages* Nauruan (official), English
*religion* Protestant 66%, Roman Catholic 33%
*literacy* 99% (1988)
*GNP* $160 million (1986); $9,091 per head (1985)
*chronology*
*1888* Annexed by Germany.
*1920* Administered by Australia, New Zealand, and UK until independence, except 1942–45, when it was occupied by Japan.
*1968* Independence achieved from Australia, New Zealand, and UK with 'special member' Commonwealth status. Hammer DeRoburt elected president.

*1976* Bernard Dowiyogo elected president.
*1978* DeRoburt re-elected.
*1986* DeRoburt briefly replaced as president by Kennan Adeang.
*1987* DeRoburt re-elected; Adeang established the Democratic Party of Nauru.
*1989* DeRoburt replaced by Kensas Aroi, who was later succeeded by Dowiyogo.
*1992* Dowiyogo re-elected.

# Nepal
Kingdom of
(*Nepal Adhirajya*)

*area* 147,181 sq km/56,850 sq mi
*capital* Katmandu
*towns* Pátan, Moráng, Bhádgáon
*physical* descends from the Himalayan mountain range in N through foothills to the river Ganges plain in S
*environment* Nepal attracts 270,000 tourists, trekkers, and mountaineers each year. An estimated 500 kg/1,100 lb of rubbish is left by each expedition trekking or climbing in the Himalayas. Since 1952 the foothills of the Himalayas have been stripped of 40% of their forest cover
*head of state* King Birendra Bir Bikram Shah Dev from 1972
*head of government* Girija Prasad Koirala from 1991
*political system* constitutional monarchy
*exports* jute, rice, timber, oilseed
*currency* Nepalese rupee
*population* (1990 est) 19,158,000 (mainly known by name of predominant clan, the Gurkhas; the Sherpas are a Buddhist minority of

NE Nepal); growth rate 2.3% p.a.
*life expectancy* men 50, women 49 (1989)
*language* Nepali (official); 20 dialects spoken
*religion* Hindu 90%; Buddhist, Muslim, Christian
*literacy* men 39%, women 12% (1985 est)
*GNP* $3.1 bn (1988); $160 per head (1986)
*chronology*
*1768* Nepal emerged as unified kingdom.
*1815–16* Anglo-Nepali 'Gurkha War'; Nepal became a British-dependent buffer state.
*1846–1951* Ruled by the Rana family.
*1923* Independence achieved from Britain.
*1951* Monarchy restored.
*1959* Constitution created elected legislature.
*1960–61* Parliament dissolved by king; political parties banned.
*1980* Constitutional referendum held following popular agitation.
*1981* Direct elections held to national assembly.
*1983* Overthrow of monarch-supported prime minister.
*1986* New assembly elections returned a majority opposed to *panchayat* system of partyless government.
*1988* Strict curbs placed on opposition activity; over 100 supporters of banned opposition party arrested; censorship imposed.
*1989* Border blockade imposed by India in treaty dispute.
*1990 Panchayat* system collapsed after mass prodemocracy demonstrations; new constitution introduced; elections set for May 1991.
*1991* Nepali Congress Party, led by Girija Prasad Koirala, won the general election.
*1992* Communists led anti-government demonstrations in Katmandu and Pátan.

# Netherlands

**Netherlands**
Kingdom of the (*Koninkrijk der Nederlanden*), popularly referred to as *Holland*

*area* 41,863 sq km/16,169 sq mi
*capital* Amsterdam
*towns* The Hague, Utrecht, Eindhoven, Maastricht, Rotterdam
*physical* flat coastal lowland; rivers Rhine, Scheldt, Maas; Frisian Islands
*territories* Aruba, Netherlands Antilles (Caribbean)
*environment* the country lies at the mouths of three of Europe's most polluted rivers, the Maas, Rhine, and Scheldt. Dutch farmers contribute to this pollution by using the world's highest quantities of nitrogen-based fertilizer per hectare/acre per year
*head of state* Queen Beatrix Wilhelmina Armgard from 1980
*head of government* Ruud Lubbers from 1989
*political system* constitutional monarchy
*exports* dairy products, flower bulbs, vegetables, petrochemicals, electronics
*currency* guilder
*population* (1990 est) 14,864,000 (including 300,000 of Dutch-Indonesian origin absorbed 1949–64 from former colonial

possessions); growth rate 0.4% p.a.
*life expectancy* men 74, women 81 (1989)
*language* Dutch
*religions* Roman Catholic 40%, Protestant 31%
*literacy* 99% (1989)
*GNP* $223 bn (1988); $13,065 per head (1987)
*chronology*
*1940–45* Occupied by Germany during World War II.
*1947* Joined Benelux customs union.
*1948* Queen Juliana succeeded Queen Wilhelmina to the throne.
*1949* Became a founding member of North Atlantic Treaty Organization (NATO).
*1953* Dykes breached by storm; nearly 2,000 people and tens of thousands of cattle died in flood.
*1958* Joined European Economic Community.
*1980* Queen Juliana abdicated in favour of her daughter Beatrix.
*1981* Opposition to cruise missiles averted their being sited on Dutch soil.
*1989* Prime Minister Ruud Lubbers resigned; new Lubbers-led coalition elected.
*1991* Treaty on political and monetary union signed by European Community (EC) members at Maastricht.
*1992* Maastricht Treaty ratified.

# New Zealand
Dominion of

*area* 268,680 sq km/103,777 sq mi
*capital* Wellington
*towns* Hamilton, Palmerston North, Christchurch, Dunedin, Auckland
*physical* comprises North Island, South Island, Stewart Island, Chatham Islands, and minor islands; mainly mountainous
*overseas territories* Tokelau (three atolls transferred 1926 from former Gilbert and Ellice Islands colony); Niue Island (one of the Cook Islands, separately administered from 1903: chief town Alafi); Cook Islands are internally self-governing but share common citizenship with New Zealand; Ross Dependency in Antarctica
*head of state* Elizabeth II from 1952, represented by governor general (Catherine Tizard from 1990)
*head of government* Jim Bolger from 1990
*political system* constitutional monarchy
*exports* lamb, beef, wool, leather, dairy products, processed foods, kiwi fruit, seeds and breeding stock, timber, paper, pulp, light aircraft
*currency* New Zealand dollar
*population* (1990 est) 3,397,000 (European, mostly British, 87%; Polynesian, mostly Maori, 12%); growth rate 0.9% p.a.
*life expectancy* men 72, women 78 (1989)
*languages* English (official), Maori

***religions*** Protestant 50%, Roman Catholic 15%
***literacy*** 99% (1989)
***GNP*** $37 bn; $11,040 per head (1988)
***chronology***
***1840*** New Zealand became a British colony.
***1907*** Created a dominion of the British Empire.
***1931*** Granted independence from Britain.
***1947*** Independence within the Commonwealth confirmed by the New Zealand parliament.
***1972*** National Party government replaced by Labour Party, with Norman Kirk as prime minister.
***1974*** Kirk died; replaced by Wallace Rowling.
***1975*** National Party returned, with Robert Muldoon as prime minister.
***1984*** Labour Party returned under David Lange.
***1985*** Non-nuclear military policy created disagreements with France and the USA.
***1987*** National Party declared support for the Labour government's non-nuclear policy. Lange re-elected. New Zealand officially classified as a 'friendly' rather than 'allied' country by the USA because of its non-nuclear military policy.
***1988*** Free-trade agreement with Australia signed.
***1989*** Lange resigned over economic differences with finance minister (he cited health reasons); replaced by Geoffrey Palmer.
***1990*** Palmer replaced by Mike Moore. Labour Party defeated by National Party in general election; Jim Bolger became prime minister.
***1991*** Formation of amalgamated Alliance Party set to challenge two-party system.
***1992*** Ban on visits by US warships lifted. Constitutional change agreed upon.

## Nicaragua
Republic of
(*República de Nicaragua*)

*area* 127,849 sq km/49,363 sq mi
*capital* Managua
*towns* León, Granada, Corinto, Puerto Cabezas, El Bluff
*physical* narrow Pacific coastal plain separated from broad Atlantic coastal plain by volcanic mountains and lakes Managua and Nicaragua
*head of state and government* Violeta Barrios de Chamorro from 1990
*political system* emergent democracy
*exports* coffee, cotton, sugar, bananas, meat
*currency* cordoba
*population* (1990) 3,606,000 (mestizo 70%, Spanish descent 15%, Indian or black 10%); growth rate 3.3% p.a.
*life expectancy* men 61, women 63 (1989)
*languages* Spanish (official), Indian, English
*religion* Roman Catholic 95%
*literacy* 66% (1986)
*GNP* $2.1 bn; $610 per head (1988)
*chronology*
*1838* Independence achieved from Spain.
*1926–1933* Occupied by US marines.
*1936* General Anastasio Somoza elected president; start of

near-dictatorial rule by Somoza family.
*1962* Sandinista National Liberation Front (FSLN) formed to fight Somoza regime.
*1979* Somoza government ousted by FSLN.
*1982* Subversive activity against the government by right-wing Contra guerrillas promoted by the USA. State of emergency declared.
*1984* The USA mined Nicaraguan harbours.
*1985* Denunciation of Sandinista government by US president Reagan. FSLN won assembly elections.
*1987* Central American peace agreement cosigned by Nicaraguan leaders.
*1988* Peace agreement failed. Nicaragua held talks with Contra rebel leaders. Hurricane left 180,000 people homeless.
*1989* Demobilization of rebels and release of former Somozan supporters; cease-fire ended.
*1990* FSLN defeated by UNO, a US-backed coalition; Violeta Barrios de Chamorro elected president. Antigovernment riots.
*1991* First presidential state visit to USA for over fifty years.
*1992* June: US aid suspended because of concern over role of Sandinista in Nicaraguan government. Sept: around 16,000 made homeless by earthquake.

# Niger
Republic of
(*République du Niger*)

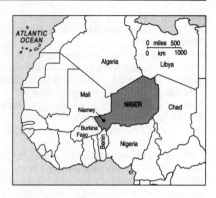

*area* 1,186,408 sq km/457,953 sq mi
*capital* Niamey
*towns* Zinder, Maradi, Tahoua
*physical* desert plains between hills in N and savanna in S; river Niger in SW, Lake Chad in SE
*head of state* Ali Saibu from 1987
*head of government* to be elected
*political system* military republic
*exports* peanuts, livestock, gum arabic, uranium
*currency* franc CFA
*population* (1990 est) 7,691,000; growth rate 2.8% p.a.
*life expectancy* men 48, women 50 (1989)
*languages* French (official), Hausa, Djerma, and other minority languages
*religions* Sunni Muslim 85%, animist 15%
*literacy* men 19%, women 9% (1985 est)
*GNP* $2.2 bn; $310 per head (1987)
*chronology*
*1960* Achieved full independence from France; Hamani Diori elected first president.
*1974* Diori ousted in army coup led by Seyni Kountché.

*1977* Cooperation agreement signed with France.
*1987* Kountché died; replaced by Col Ali Saibu.
*1989* Saibu elected president without opposition.
*1990* Multiparty politics promised.
*1991* Saibu stripped of executive powers; transitional government formed.
*1992* Transitional government collapsed. Constitutional change allowing for multiparty politics approved in referendum.

# Nigeria
Federal Republic of

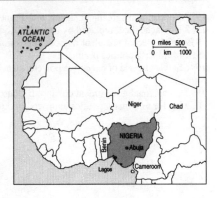

*area* 923,773 sq km/356,576 sq mi
*capital* Abuja
*towns* Ibadan, Ogbomosho, Kano, Lagos, Port Harcourt, Warri
*physical* arid savanna in N; tropical rainforest in S, with mangrove swamps along the coast; river Niger forms wide delta; mountains in SE
*environment* toxic waste from northern industrialized countries has been dumped in Nigeria
*head of state and government* Ibrahim Babangida from 1985
*political system* military republic pending promised elections
*exports* petroleum (largest oil resources in Africa), cocoa, peanuts, palm oil (Africa's largest producer), cotton, rubber, tin
*currency* naira
*population* (1991) 88,514,500 (Yoruba in W, Ibo in E, and Hausa–Fulani in N); growth rate 3.3% p.a.
*life expectancy* men 47, women 49 (1989)
*languages* English (official), Hausa, Ibo, Yoruba
*religions* Sunni Muslim 50% (in N), Christian 40% (in S), local religions 10%
*literacy* men 54%, women 31% (1985 est)
*GNP* $78 bn (1987); $790 per head (1984)

*chronology*
*1914* N Nigeria and S Nigeria united to become Britain's largest African colony.
*1954* Nigeria became a federation.
*1960* Independence achieved from Britain within the Commonwealth.
*1963* Became a republic, with Nnamdi Azikiwe as president.
*1966* Military coup, followed by a counter-coup led by General Yakubu Gowon. Slaughter of many members of the Ibo tribe in north.
*1967* Conflict over oil revenues led to declaration of an independent Ibo state of Biafra and outbreak of civil war.
*1970* Surrender of Biafra and end of civil war.
*1975* Gowon ousted in military coup; second coup put General Olusegun Obasanjo in power.
*1979* Shehu Shagari became civilian president.
*1983* Shagari's government overthrown in coup by Maj-Gen Muhammadu Buhari.
*1985* Buhari replaced in a bloodless coup led by Maj-General Ibrahim Babangida.
*1989* Two new political parties approved. Babangida promised a return to pluralist politics; date set for 1992.
*1991* Nine new states created. Babangida confirmed his commitment to democratic rule for 1992.
*1992* Multiparty elections won by Babangida's Social Democratic Party (SDP). Primary elections to be introduced; transition to civilian rule delayed.

# Norway
Kingdom of
(*Kongeriket Norge*)

*area* 387,000 sq km/149,421 sq mi (includes Svalbard and Jan Mayen)
*capital* Oslo
*towns* Bergen, Trondheim, Stavanger
*physical* mountainous with fertile valleys and deeply indented coast; forests cover 25%; extends N of Arctic Circle
*territories* dependencies in the Arctic (Svalbard and Jan Mayen) and in Antarctica (Bouvet and Peter I Island, and Queen Maud Land)
*environment* an estimated 80% of the lakes and streams in the southern half of the country have been severely acidified by acid rain
*head of state* Harald V from 1991
*head of government* Gro Harlem Brundtland from 1990
*political system* constitutional monarchy
*exports* petrochemicals from North Sea oil and gas, paper, wood pulp, furniture, iron ore and other minerals, high-tech goods, fish
*currency* krone
*population* (1990 est) 4,214,000; growth rate 0.3% p.a.
*life expectancy* men 73, women 80 (1989)
*languages* Norwegian (official); there are Saami- (Lapp-) and Finnish-speaking minorities
*religion* Evangelical Lutheran (endowed by state) 94%
*literacy* 100% (1989)

**GNP** $89 bn (1988); $13,790 per head (1984)

*chronology*

*1814* Became independent from Denmark; ceded to Sweden.
*1905* Links with Sweden ended; full independence achieved.
*1940–45* Occupied by Germany.
*1949* Joined North Atlantic Treaty Organization (NATO).
*1952* Joined Nordic Council.
*1957* King Haakon VII succeeded by his son Olaf V.
*1960* Joined European Free Trade Association (EFTA).
*1972* Accepted into membership of European Economic Community; application withdrawn after a referendum.
*1988* Gro Harlem Brundtland awarded Third World Prize.
*1989* Jan P Syse became prime minister.
*1990* Brundtland returned to power.
*1991* King Olaf V died; succeeded by his son Harald V.
*1992* Defied whaling ban to resume whaling industry. Brundtland relinquished leadership of the Labour Party. Formal application made for EC membership.

# Oman
Sultanate of
(*Saltanat 'Uman*)

*area* 272,000 sq km/105,000 sq mi
*capital* Muscat
*towns* Salalah, Nizwa
*physical* mountains to N and S of a high arid plateau; fertile coastal strips NW and S
*head of state and government* Qaboos bin Said from 1970
*political system* absolute monarchy
*exports* oil, dates, silverware, copper
*currency* rial Omani
*population* (1990 est) 1,305,000; growth rate 3.0% p.a.
*life expectancy* men 55, women 58 (1989)
*languages* Arabic (official), English, Urdu, other Indian languages
*religions* Ibadhi Muslim 75%, Sunni Muslim, Shi'ite Muslim, Hindu
*literacy* 20% (1989)
*GNP* $7.5 bn (1987); $5,070 per head (1988)
*chronology*
*1951* The Sultanate of Muscat and Oman achieved full independence from Britain. Treaty of Friendship with Britain signed.
*1970* After 38 years' rule, Sultan Said bin Taimur replaced in coup by his son Qaboos bin Said. Name changed to Sultanate of Oman.

*1975* Left-wing rebels in south defeated.
*1982* Memorandum of Understanding with UK signed, providing for regular consultation on international issues.
*1985* Diplomatic ties established with USSR.
*1991* Sent troops to Operation Desert Storm, as part of coalition opposing Iraq's occupation of Kuwait.

## Pakistan
Islamic Republic of

*area* 796,100 sq km/307,295 sq mi; one-third of Kashmir under Pakistani control
*capital* Islamabad
*towns* Karachi, Lahore, Rawalpindi, Peshawar
*physical* fertile Indus plain in E, Baluchistan plateau in W, mountains in N and NW
*environment* about 68% of irrigated land is waterlogged or suffering from salinization
*head of state* Ghulam Ishaq Khan from 1988
*head of government* Nawaz Sharif from 1990
*political system* emergent democracy
*exports* cotton textiles, rice, leather, carpets
*currency* Pakistan rupee
*population* (1990 est) 113,163,000 (Punjabi 66%, Sindhi 13%); growth rate 3.1% p.a.
*life expectancy* men 54, women 55 (1989)
*languages* Urdu and English (official); Punjabi, Sindhi, Pashto, Baluchi, other local dialects
*religion* Sunni Muslim 75%, Shi'ite Muslim 20%, Hindu 4%

*literacy* men 40%, women 19% (1985 est)
*GDP* $39 bn (1988); $360 per head (1984)
*chronology*
*1947* Independence achieved from Britain; Pakistan formed following partition of British India.
*1956* Proclaimed a republic.
*1958* Military rule imposed by General Ayub Khan.
*1969* Power transferred to General Yahya Khan.
*1971* Secession of East Pakistan (Bangladesh). After civil war, power transferred to Zulfiqar Ali Bhutto.
*1977* Bhutto overthrown in military coup by General Zia ul-Haq; martial law imposed.
*1979* Bhutto executed.
*1981* Opposition Movement for the Restoration of Democracy formed. Islamization process pushed forward.
*1985* Nonparty elections held; amended constitution adopted; martial law and ban on political parties lifted.
*1986* Agitation for free elections launched by Benazir Bhutto.
*1988* Zia introduced Islamic legal code, the Shari'a. He was killed in a military plane crash Aug. Benazir Bhutto elected prime minister Nov.
*1989* Pakistan rejoined the Commonwealth.
*1990* Army mobilized in support of Muslim separatists in Indian Kashmir. Bhutto dismissed on charges of incompetence and corruption. Islamic Democratic Alliance (IDA), led by Nawaz Sharif, won Oct general election.
*1991* Shari'a bill enacted; privatization and economic deregulation programme launched.
*1992* Sept: Floods devastated north of country. Oct: Pakistan elected to UN Security Council 1993–95.

# Panama
Republic of
(*República de Panamá*)

*area* 77,100 sq km/29,768 sq mi
*capital* Panamá (Panama City)
*towns* Cristóbal, Balboa, Colón, David
*physical* coastal plains and mountainous interior; tropical rainforest in E and NW; Pearl Islands in Gulf of Panama
*head of state and government* Guillermo Endara from 1989
*political system* emergent democratic republic
*exports* bananas, petroleum products, copper, shrimps, sugar
*currency* balboa
*population* (1990 est) 2,423,000 (mestizo, or mixed race, 70%, West Indian 14%, European descent 10%, Indian (Cuna, Choco, Guayami) 6%); growth rate 2.2% p.a.
*life expectancy* men 71, women 75 (1989)
*languages* Spanish (official), English
*religions* Roman Catholic 93%, Protestant 6%
*literacy* 87% (1989)
*GNP* $4.2 bn (1988); $1,970 per head (1984)
*chronology*
*1821* Achieved independence from Spain; joined the confederacy of

Gran Colombia.

*1903* Full independence achieved on separation from Colombia.

*1974* Agreement to negotiate full transfer of the Panama Canal from the USA to Panama.

*1977* USA–Panama treaties transferred the canal to Panama, effective from 1999, with the USA guaranteeing its protection and an annual payment.

*1984* Nicolás Ardito Barletta elected president.

*1985* Barletta resigned; replaced by Eric Arturo del Valle.

*1987* General Noriega (head of the National Guard and effective ruler of Panama) resisted calls for his removal, despite suspension of US military and economic aid.

*1988* Del Valle replaced by Manuel Solis Palma. Noriega, charged with drug smuggling by the USA, declared a state of emergency.

*1989* Opposition won election; Noriega declared results invalid; Francisco Rodríguez sworn in as president. Coup attempt against Noriega failed; Noriega declared head of government by assembly. 'State of war' with the USA announced. US invasion deposed Noriega; Guillermo Endara installed as president; Noriega sought asylum in Vatican embassy; later surrendered and taken to US for trial.

*1991* Attempted antigovernment coup foiled. Army abolished.

*1992* Noriega found guilty of drug offences. Referendum voted down overwhelmingly government's constitutional changes, including abolition of a standing army.

# Papua New Guinea

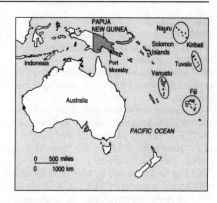

*area* 462,840 sq km/178,656 sq mi
*capital* Port Moresby (on E New Guinea)
*towns* Lae, Rabaul, Madang
*physical* mountainous; includes tropical islands of New Ireland, New Britain, and Bougainville; Admiralty Islands, D'Entrecasteaux Islands, and Louisiade Archipelago
*head of state* Elizabeth II, represented by governor general
*head of government* Paias Wingti from 1992
*political system* liberal democracy
*exports* copra, coconut oil, palm oil, tea, copper, gold, coffee
*currency* kina
*population* (1989 est) 3,613,000 (Papuans, Melanesians, Negritos, various minorities); growth rate 2.6% p.a.
*life expectancy* men 53, women 54 (1987)
*languages* English (official); pidgin English, 715 local languages
*religions* Protestant 63%, Roman Catholic 31%, local faiths
*literacy* men 55%, women 36% (1985 est)
*GNP* $2.5 bn; $730 per head (1987)
*chronology*
*1883* Annexed by Queensland; became the Australian Territory of Papua.

*1884* NE New Guinea annexed by Germany; SE claimed by Britain.
*1914* NE New Guinea occupied by Australia.
*1921–42* Held as a League of Nations mandate.
*1942–45* Occupied by Japan.
*1975* Independence achieved from Australia, within the Commonwealth, with Michael Somare as prime minister.
*1980* Julius Chan became prime minister.
*1982* Somare returned to power.
*1985* Somare challenged by deputy prime minister, Paias Wingti, who later formed a five-party coalition government.
*1988* Wingti defeated on no-confidence vote and replaced by Rabbie Namaliu, who established a six-party coalition government.
*1989* State of emergency imposed on Bougainville in response to separatist violence.
*1991* Peace accord signed with Bougainville secessionists. Economic boom as gold production doubled. Wiwa Korowi elected as new governor general. Deputy prime minister, Ted Diro, resigned, having been found guilty of corruption.
*1992* April: killings by outlawed Bougainville secessionists reported. July: Wingti elected premier.

## Paraguay
Republic of
(*República del Paraguay*)

***area*** 406,752 sq km/157,006 sq mi
***capital*** Asunción
***towns*** Puerto Presidente Stroessner, Pedro Juan Caballero, Concepción
***physical*** low marshy plain and marshlands; divided by Paraguay River; Paraná River forms SE boundary
***head of state and government*** General Andrés Rodríguez from 1989
***political system*** emergent democratic republic
***exports*** cotton, soya beans, timber, vegetable oil, maté
***currency*** guaraní
***population*** (1990 est) 4,660,000 (95% mixed Guarani Indian–Spanish descent); growth rate 3.0% p.a.
***life expectancy*** men 67, women 72 (1989)
***languages*** Spanish 6% (official), Guarani 90%
***religion*** Roman Catholic 97%
***literacy*** men 91%, women 85% (1985 est)
***GNP*** $7.4 bn; $1,000 per head (1987)

*chronology*
*1811* Independence achieved from Spain.
*1865–70* War with Argentina, Brazil, and Uruguay; less than half the population survived and much territory was lost.
*1932–35* Territory won from Bolivia during the Chaco War.
*1940–48* Presidency of General Higinio Morínigo.
*1948–54* Political instability; six different presidents.
*1954* General Alfredo Stroessner seized power.
*1989* Stroessner ousted in coup led by General Andrés Rodríguez. Rodríguez elected president; Colorado Party won the congressional elections.
*1991* Colorado Party successful in assembly elections.

## Peru
Republic of
(*República del Perú*)

*area* 1,285,200 sq km/496,216 sq mi
*capital* Lima, including port of Callao
*towns* Arequipa, Iquitos, Chiclayo, Trujillo
*physical* Andes mountains NW–SE cover 27% of Peru, separating Amazon rainforest in NE from coastal plain in W; desert along coast
*environment* an estimated 38% of the 8,000 sq km/3,100 sq mi of coastal lands under irrigation is either waterlogged or suffering from saline water. Only half the population has access to clean drinking water
*head of state and government* Alberto Fujimori from 1990
*political system* democratic republic
*exports* coca, coffee, alpaca, llama and vicuña wool, fish meal, lead (largest producer in South America), copper, iron, oil
*currency* new sol
*population* (1990 est) 21,904,000 (Indian, 46%; mixed Spanish–Indian descent 43%); growth rate 2.6% p.a.

*life expectancy* men 61, women 66
*languages* Spanish 68%, Quechua 27% (both official), Aymara 3%
*religion* Roman Catholic 90%
*literacy* men 91%, women 78% (1985 est)
*GNP* $19.6 bn (1988); $940 per head (1984)
*chronology*
*1824* Independence achieved from Spain.
*1849–74* Some 80,000–100,000 Chinese labourers arrived in Peru to fill menial jobs such as collecting guano.
*1902* Boundary dispute with Bolivia settled.
*1927* Boundary dispute with Colombia settled.
*1942* Boundary dispute with Ecuador settled.
*1948* Army coup, led by General Manuel Odría, installed a military government.
*1963* Return to civilian rule, with Fernando Belaúnde Terry as president.
*1968* Return of military government in a bloodless coup by General Juan Velasco Alvarado.
*1975* Velasco replaced, in a bloodless coup, by General Morales Bermúdez.
*1980* Return to civilian rule, with Fernando Belaúnde as president.
*1981* Boundary dispute with Ecuador renewed.
*1985* Belaúnde succeeded by Social Democrat Alan García Pérez.
*1988* García pressured to seek help from the International Monetary Fund. (IMF)
*1989* Mario Vargas Llosa entered presidential race; his Democratic Front won municipal elections Nov.
*1990* Alberto Fujimori defeated Vargas Llosa in presidential elections. Assassination attempt on president failed.
*1992* April: Fujimori sided with army to avert coup, announcing crackdown on rebels and drug traffickers. USA suspended humanitarian aid. Sendero Luminoso ('Shining Path') terrorists continued campaign of violence. May: Fujimori promised return to democracy. Oct: Sendero Luminoso leader, Abimael Guzman Reynoso, arrested received life sentence. Nov: anti-government coup foiled; single-chamber legislature replaced two-chamber system.

# Philippines
Republic of the
(*Republika ng Pilipinas*)

*area* 300,000 sq km/115,800 sq mi
*capital* Manila (on Luzon)
*towns* Quezon City (Luzon), Zamboanga (Mindanao), Cebu, Davao (on Mindanao), Iloilo
*physical* comprises over 7,000 islands; volcanic mountain ranges traverse main chain N–S; 50% still forested. The largest islands are Luzon 108,172 sq km/41,754 sq mi and Mindanao 94,227 sq km/36,372 sq mi; others include Samar, Negros, Palawan, Panay, Mindoro, Leyte, Cebu, and the Sulu group
*environment* cleared for timber, tannin, and the creation of fish ponds, the mangrove forest was reduced from 5,000 sq km/1,930 sq mi to 380 sq km/146 sq mi between 1920 and 1988
*head of state and government* Fidel Ramos from 1992
*political system* emergent democracy
*exports* sugar, copra (world's largest producer) and coconut oil, timber, copper concentrates, electronics, clothing
*currency* peso
*population* (1990 est) 66,647,000 (93% Malaysian); growth rate 2.4% p.a.

***life expectancy*** men 63, women 69 (1989)
***languages*** Tagalog (Filipino, official); English and Spanish
***religions*** Roman Catholic 84%, Protestant 9%, Muslim 5%
***literacy*** 88% (1989)
***GNP*** $38.2 bn; $667 per head (1988)
***chronology***
***1542*** Named the Philippines (Filipinas) by Spanish explorers.
***1565*** Conquered by Spain.
***1898*** Ceded to the USA after Spanish–American War.
***1935*** Granted internal self-government.
***1942–45*** Occupied by Japan.
***1946*** Independence achieved from USA.
***1965*** Ferdinand Marcos elected president.
***1983*** Opposition leader Benigno Aquino murdered by military guard.
***1986*** Marcos overthrown by Corazon Aquino's People's Power movement.
***1987*** 'Freedom constitution' adopted, giving Aquino mandate to rule until June 1992; People's Power won majority in congressional elections. Attempted right-wing coup suppressed. Communist guerrillas active. Government in rightward swing.
***1988*** Land Reform Act gave favourable compensation to holders of large estates.
***1989*** Referendum on southern autonomy failed; Marcos died in exile; Aquino refused his burial in Philippines. Sixth coup attempt suppressed with US aid; Aquino declared state of emergency.
***1990*** Seventh coup attempt survived by President Aquino.
***1991*** June: eruption of Mount Pinatubo, hundreds killed. USA agreed to give up Clark Field airbase but keep Subic Bay naval base for ten more years. Sept: Philippines Senate voted to urge withdrawal of all US forces. US renewal of Subic Bay lease rejected. Nov: Imelda Marcos returned.
***1992*** Fidel Ramos elected to replace Aquino.

## Poland
Republic of
(*Polska Rzeczpospolita*)

*area* 127,886 sq km/49,325 sq mi
*capital* Warsaw
*towns* Lódź, Kraków, Wroclaw, Poznań, Katowice, Gdańsk
*physical* part of the great plain of Europe; Vistula, Oder, and Neisse rivers; Sudeten, Tatra, and Carpathian mountains on S frontier
*environment* atmospheric pollution derived from coal (producing 90% of the country's electricity), toxic waste from industry, and lack of sewage treatment have resulted in the designation of 27 ecologically endangered areas. Half the country's lakes have been seriously contaminated and three-quarters of its drinking water does not meet official health standards
*head of state* Lech Walesa from 1990
*head of government* Hanna Suchocka from 1992
*political system* emergent democratic republic
*exports* coal, softwood timber, chemicals, machinery, ships, vehicles, meat, copper (Europe's largest producer)
*currency* zloty
*population* (1990 est) 38,363,000; growth rate 0.6% p.a.
*life expectancy* men 66, women 74 (1989)

***languages*** Polish (official), German
***religion*** Roman Catholic 95%
***literacy*** 98% (1989)
***GNP*** $276 billion (1988); $2,000 per head (1986)
***chronology***
***1918*** Poland revived as independent republic.
***1939*** German invasion and occupation.
***1944*** Germans driven out by Soviet forces.
***1945*** Polish boundaries redrawn at Potsdam Conference.
***1947*** Communist people's republic proclaimed.
***1956*** Poznań riots. Wladyslav Gomulka installed as Polish United Workers' Party (PUWP) leader.
***1970*** Gomulka replaced by Edward Gierek after Gdańsk riots.
***1980*** Solidarity emerged as a free trade union following Gdańsk disturbances.
***1981–83*** Martial law imposed by General Wojciech Jaruzelski.
***1985*** Zbigniew Messner became prime minister.
***1988*** Solidarity-led strikes and demonstrations called off after pay increases. Messner replaced by reformist Mieczyslaw Rakowski.
***1989*** Solidarity relegalized. April: new 'socialist pluralist' constitution formed. June: widespread success for Solidarity in assembly elections, the first open elections in 40 years. July: Jaruzelski elected president. Sept: 'Grand coalition', first non-Communist government since World War II formed; economic restructuring undertaken on free-market lines; W Europe and US create $1 billion aid package.
***1990*** Jan: PUWP dissolved; replaced by Social Democratic Party and breakaway Union of Social Democrats. Lech Walesa elected president; Dec: prime minister Mazowiecki resigned.
***1991*** Oct: Multiparty general election produced inconclusive result. Five-party centre-right coalition formed under Jan Olszewski. Treaty signed agreeing to complete withdrawal of Soviet troops.
***1992*** June: Olszewski ousted on vote of no confidence; succeeded by Waldemar Pawlak. July: Hanna Suchocka replaced Pawlak as Poland's first woman prime minister.
***1993*** March: 14% of workforce (2.6 million) out of work.

# Portugal
Republic of
(*República Portuguesa*)

*area* 92,000 sq km/35,521 sq mi (including the Azores and Madeira)
*capital* Lisbon
*towns* Coimbra, Pôrto, Setúbal
*physical* mountainous in N, plains in S
*head of state* Mario Alberto Nobre Lopes Soares from 1986
*head of government* Aníbal Cavaco Silva from 1985
*political system* democratic republic
*exports* wine, olive oil, resin, cork, sardines, textiles, clothing, pottery, pulpwood
*currency* escudo
*population* (1990 est) 10,528,000; growth rate 0.5% p.a.
*life expectancy* men 71, women 78 (1989)
*language* Portuguese
*religion* Roman Catholic 97%
*literacy* men 89%, women 80% (1985)
*GNP* $33.5 bn (1987); $2,970 per head (1986)
*chronology*
*1928–68* Military dictatorship under António de Oliveira Salazar.
*1968* Salazar succeeded by Marcello Caetano.

*1974* Caetano removed in military coup led by General Antonio Ribeiro de Spínola. Spínola replaced by General Francisco da Costa Gomes.
*1975* African colonies became independent.
*1976* New constitution, providing for return to civilian rule, adopted. Minority government appointed, led by Socialist Party leader Mario Soares.
*1978* Soares resigned.
*1980* Francisco Balsemão formed centre-party coalition after two years of political instability.
*1982* Draft of new constitution approved, reducing powers of presidency.
*1983* Centre-left coalition government formed.
*1985* Aníbal Cavaco Silva became prime minister.
*1986* Mario Soares elected first civilian president in 60 years. Portugal joined European Community.
*1988* Portugal joined Western European Union.
*1989* Constitution amended to allow major state enterprises to be denationalized.
*1991* Mario Soares re-elected president; Social Democrat (PSD) majority slightly reduced in assembly elections.

**Qatar**
State of
(*Dawlat Qatar*)

*area* 11,400 sq km/4,402 sq mi
*capital* Doha
*town* Dukhan
*physical* mostly flat desert with salt flats in S
*head of state and government* Sheik Khalifa bin Hamad al-Thani from 1972
*political system* absolute monarchy
*exports* oil, natural gas, petrochemicals, fertilizers, iron, steel
*currency* riyal
*population* (1990 est) 498,000 (half in Doha; Arab 40%, Indian 18%, Pakistani 18%); growth rate 3.7% p.a.
*life expectancy* men 68, women 72 (1989)
*languages* Arabic (official), English
*religion* Sunni Muslim 95%
*literacy* 60% (1987)
*GNP* $5.9 bn (1983); $35,000 per head
*chronology*
*1916* Qatar became a British protectorate.
*1939* Oil was discovered in W Qatar.
*1949* Qatar began exporting oil.

*1970* Constitution adopted, confirming the emirate as an absolute monarchy
*1971* Independence achieved from Britain
*1972* Emir Sheik Ahmad replaced in bloodless coup by his cousin, Crown Prince Sheik Khalifa
*1991* Forces joined UN coalition in Gulf War against Iraq.

# Romania

***area*** 237,500 sq km/91,699 sq mi
***capital*** Bucharest
***towns*** Braşov, Timişoara, Cluj–Napoca, Iaşi, Galaţi, Constanta, Braila
***physical*** mountains surrounding a plateau, with river plains S and E
***environment*** although sulphur-dioxide levels are low, only 20% of the country's rivers can provide drinkable water
***head of state*** Ion Iliescu from 1989
***head of government*** Theodor Stolojan from 1991
***political system*** emergent democratic republic
***exports*** petroleum products and oilfield equipment, electrical goods, cars, cereals
***currency*** leu
***population*** (1990 est) 23,269,000 (Romanians 89%, Hungarians 7.9%, Germans 1.6%); growth rate 0.5% p.a.
***life expectancy*** men 67, women 73 (1989)
***languages*** Romanian (official), Hungarian, German
***religions*** Romanian Orthodox 80%, Roman Catholic 6%
***literacy*** 98% (1988)

**GNP** $151 bn (1988); $6,400 per head
*recent chronology*
**1944** Pro-Nazi Antonescu government overthrown.
**1945** Communist-dominated government appointed.
**1947** Boundaries redrawn. King Michael abdicated and People's Republic proclaimed.
**1949** New Soviet-style constitution adopted. Joined Comecon.
**1952** Second new Soviet-style constitution.
**1955** Romania joined Warsaw Pact.
**1958** Soviet occupation forces removed.
**1965** New constitution adopted.
**1974** Ceauşescu created president.
**1985–86** Winters of austerity and power cuts.
**1987** Workers demonstrated against austerity programme.
**1988–89** Relations with Hungary deteriorated over 'systematization programme'.
**1989** Announcement that all foreign debt paid off. Razing of villages and building of monuments to Ceauşescu. Communist orthodoxy reaffirmed; demonstrations violently suppressed; massacre in Timisoara. Army joined uprising; heavy fighting; bloody overthrow of Ceauşescu regime in 'Christmas Revolution'; Ceauşescu and wife tried and executed; estimated 10,000 dead in civil warfare. Power assumed by new National Salvation Front, headed by Ion Iliescu.
**1990** Securitate secret police replaced by new Romanian Intelligence Service (RIS); religious practices resumed; mounting strikes and protests against effects of market economy.
**1991** April: treaty on cooperation and good neighbourliness signed with USSR. Aug: privatization law passed. Sept: prime minister Petre Roman resigned following riots; succeeded by Theodor Stolojan heading a new cross-party coalition government. Dec: new constitution endorsed by referendum.
**1992** Iliescu re-elected in presidential runoff.

# Russian Federation

**Russian Federation**
formerly (until 1991)
Russian Soviet
Federal Socialist
Republic (RSFSR)

*area* 17,075,500 sq km/6,591,100 sq mi
*capital* Moscow
*towns* St Petersburg (Leningrad), Nizhny-Novgorod (Gorky), Rostov-on-Don, Samara (Kuibyshev), Tver (Kalinin), Volgograd
*physical* fertile Black Earth district; extensive forests; the Ural Mountains with large mineral resources
*head of state* Boris Yeltsin from 1990/91
*head of government* Viktor Chernomyrdin from 1992
*political system* emergent democracy
*products* iron ore, coal, oil, gold, platinum, and other minerals, agricultural produce
*currency* rouble
*population* (1990) 148,000,000 (82% Russian, Tatar 4%, Ukrainian 3%, Chuvash 1%)
*language* Great Russian
*religion* traditionally Russian Orthodox
*recent chronology*
*1945* Became a founding member of United Nations.
*1988* Aug: Democratic Union formed in Moscow as political party

opposed to totalitarianism. Oct: Russian-language demonstrations in Leningrad; Tsarist flag raised.

*1989* March: Boris Yeltsin elected to USSR Congress of People's Deputies. Sept: conservative-nationalist Russian United Workers' Front established in Sverdlovsk.

*1990* May: Yeltsin narrowly elected RSFSR president by Russian parliament. June: economic and political sovereignty declared; Ivan Silaev became Russian prime minister. July: Yeltsin resigned his party membership. Dec: rationing introduced in some cities; private land ownership allowed.

*1991* March: Yeltsin secured the support of Congress of Peoples' Deputies for direct election of an executive president. June: Yeltsin elected president under a liberal-radical banner. Aug: Yeltsin stood out against abortive anti-Gorbachev coup, emerging as key power-broker within Soviet Union; pre-revolutionary flag restored. Sept: Silaev resigned as Russian premier. Nov: Yeltsin also named as prime minister; Communist Party of the Soviet Union and Russian Communist Party banned; Yeltsin's goverment gained control of Russia's economic assets and armed forces. Dec: Yeltsin negotiated formation of new confederal Commonwealth of Independent States (CIS); Russia admitted into United Nations (UN); independence recognized by USA and European Community (EC).

*1992* Jan: admitted into Conference on Security and Cooperation in Europe (CSCE); assumed former USSR's permanent seat on UN Security Council; prices freed. Feb: demonstrations in Moscow and other cities as living standards plummeted. March: 18 out of 20 republics signed treaty agreeing to remain within loose Russian Federation; Tatarstan and Checheno-Ingush refused to sign. Aug: agreement with Ukraine on joint control of Black Sea fleet. Dec: Victor Chernomyrdin elected prime minister; new constitution agreed in referendum. START II arms-reduction agreement signed with USA.

*1993* March: Congress of People's Deputies attempted to limit Yeltsin's powers to rule by decree and cancel referendum due 25 April. Yeltsin declared temporary presidential 'special rule' pending holding of referendum. Results of referendum showed Russians in favour of Yeltsin's reforms.

# Rwanda
Republic of
(*Republika y'u Rwanda*)

***area*** 26,338 sq km/10,173 sq mi
***capital*** Kigali
***towns*** Butare, Ruhengeri
***physical*** high savanna and hills, with volcanic mountains in NW
***head of state and government*** Maj-Gen Juvenal Habyarimana from 1973
***political system*** one-party military republic
***exports*** coffee, tea, pyrethrum
***currency*** franc
***population*** (1990 est) 7,603,000 (Hutu 90%, Tutsi 9%, Twa 1%); growth rate 3.3% p.a.
***life expectancy*** men 49, women 53 (1989)
***languages*** Kinyarwanda, French (official); Kiswahili
***religions*** Roman Catholic 54%, animist 23%, Protestant 12%; Muslim 9%
***literacy*** men 50% (1989)
***GNP*** $2.3 bn (1987); $323 per head (1986)
***chronology***
*1916* Belgian troops occupied Rwanda; League of Nations mandated Rwanda and Burundi to Belgium as Territory of Ruanda-Urundi.
*1959* Interethnic warfare between Hutu and Tutsi.

*1962* Independence from Belgium achieved, with Grégoire Kayibanda as president.
*1972* Renewal of interethnic fighting.
*1973* Kayibanda ousted in a military coup led by Maj-Gen Juvenal Habyarimana.
*1978* New constitution approved; Rwanda remained a military-controlled state.
*1980* Civilian rule adopted.
*1988* Refugees from Burundi massacres streamed into Rwanda.
*1990* Government attacked by Rwandan Patriotic Front (FPR), a Tutsi military-political organization based in Uganda. Constitutional reforms promised.
*1992* Peace accord with FPR.
*1993* Power-sharing agreement with government repudiated by FPR.

# St Christopher (St Kitts)–Nevis
Federation of

*area* 269 sq km/104 sq mi (St Christopher 176 sq km/68 sq mi, Nevis 93 sq km/36 sq mi)
*capital* Basseterre (on St Christopher)
*towns* Charlestown (largest on Nevis)
*physical* both islands are volcanic
*head of state* Elizabeth II from 1983, represented by governor general
*head of government* Kennedy Simmonds from 1980
*political system* federal constitutional monarchy
*exports* sugar, molasses, electronics, clothing
*currency* Eastern Caribbean dollar
*population* (1990 est) 45,800; growth rate 0.2% p.a.
*life expectancy* men 69, women 72
*language* English
*religion* Anglican 36%, Methodist 32%, other Protestant 8%, Roman Catholic 10% (1985 est)
*literacy* 90% (1987)
*GNP* $40 million (1983); $870 per head
*chronology*
*1871–1956* Part of the Leeward Islands Federation.
*1958–62* Part of the Federation of the West Indies.

*1967* St Christopher, Nevis, and Anguilla achieved internal self-government, within the British Commonwealth, with Robert Bradshaw, Labour Party leader, as prime minister.
*1971* Anguilla returned to being a British dependency.
*1978* Bradshaw died; succeeded by Paul Southwell.
*1979* Southwell died; succeeded by Lee L Moore.
*1980* Coalition government led by Kennedy Simmonds.
*1983* Full independence achieved within the Commonwealth.
*1984* Coalition government re-elected.
*1989* Prime Minister Simmonds won a third successive term.

# St Lucia

*area* 617 sq km/238 sq mi
*capital* Castries
*towns* Vieux-Fort, Soufrière
*physical* mountainous island with fertile valleys; mainly tropical forest
*head of state* Elizabeth II from 1979, represented by governor general
*head of government* John Compton from 1982
*political system* constitutional monarchy
*exports* coconut oil, bananas, cocoa, copra
*currency* Eastern Caribbean dollar
*population* (1990 est) 153,000; growth rate 2.8% p.a.
*life expectancy* men 68, women 73 (1989)
*languages* English; French patois
*religion* Roman Catholic 90%
*literacy* 78% (1989)
*GNP* $166 million; $1,370 per head (1987)
*chronology*
*1814* Became a British crown colony following Treaty of Paris.
*1967* Acquired internal self-government as a West Indies associated state.

*1979* Independence achieved from Britain, within the Commonwealth. John Compton, leader of the United Workers' Party (UWP), became prime minister. Allan Louisy, leader of the St Lucia Labour Party (SLP), replaced Compton as prime minister.
*1981* Louisy resigned; replaced by Winston Cenac.
*1982* Compton returned to power at the head of a UWP government.
*1987* Compton re-elected with reduced majority.
*1991* Integration with Windward Islands proposed.
*1992* UWP won general election.

# St Vincent and the Grenadines

*area* 388 sq km/150 sq mi, including islets of the Northern Grenadines 43 sq km/17 sq mi
*capital* Kingstown
*towns* Georgetown, Chateaubelair
*physical* volcanic mountains, thickly forested
*head of state* Elizabeth II from 1979, represented by governor general
*head of government* James Mitchell from 1984
*political system* constitutional monarchy
*exports* bananas, taros, sweet potatoes, arrowroot, copra
*currency* Eastern Caribbean dollar
*population* (1990 est) 106,000; growth rate –4% p.a.
*life expectancy* men 69, women 74 (1989)
*languages* English; French patois
*religions* Anglican 47%, Methodist 28%, Roman Catholic 13%
*literacy* 85% (1989)
*GNP* $188 million; $1,070 per head (1987)
*chronology*
*1783* Became a British crown colony.
*1958–62* Part of the West Indies Federation.

*1969* Achieved internal self-government.
*1979* Achieved full independence from Britain, within the Commonwealth, with Milton Cato as prime minister.
*1984* James Mitchell replaced Cato as prime minister.
*1989* Mitchell decisively re-elected.
*1991* Integration with Windward Islands proposed.

# Samoa, Western
Independent State of
(*Samoa i Sisifo*)

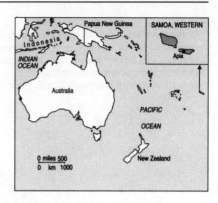

*area* 2,830 sq km/1,093 sq mi
*capital* Apia (on Upolu island)
*physical* comprises South Pacific islands of Savai'i and Upolu, with two smaller tropical islands and islets; mountain ranges on main islands
*head of state* King Malietoa Tanumafili II from 1962
*head of government* Tofilau Eti Alesana from 1988
*political system* liberal democracy
*exports* coconut oil, copra, cocoa, fruit juice, cigarettes, timber
*currency* talà
*population* (1989) 169,000; growth rate 1.1% p.a.
*life expectancy* men 64, women 69 (1989)
*languages* English, Samoan (official)
*religions* Protestant 70%, Roman Catholic 20%
*literacy* 90% (1989)
*GNP* $110 million (1987); $520 per head
*chronology*
*1899–1914* German protectorate.
*1920–61* Administered by New Zealand.
*1959* Local government elected.
*1961* Referendum favoured independence.

*1962* Independence achieved within the Commonwealth, with Fiame Mata Afa Mulinu'u as prime minister.
*1975* Mata Afa died.
*1976* Tupuola Taisi Efi became first nonroyal prime minister.
*1982* Va'ai Kolone became prime minister; replaced by Tupuola Efi. Assembly failed to approve budget; Tupuola Efi resigned; replaced by Tofilau Eti Alesana.
*1985* Tofilau Eti resigned; head of state invited Va'ai Kolone to lead the government.
*1988* Elections produced a hung parliament, with first Tupuola Efi as prime minister and then Tofilau Eti Alesana.
*1990* Universal adult suffrage introduced.
*1991* Tofilau Eti Alesana re-elected. Fiame Naome became first woman in cabinet.

# San Marino
Republic of
(*Repubblica di San Marino*)

*area* 61 sq km/24 sq mi
*capital* San Marino
*towns* Serravalle (industrial centre)
*physical* on the slope of Mount Titano
*heads of state and government* two captains regent, elected for a six-month period
*political system* direct democracy
*exports* wine, ceramics, paint, chemicals, building stone
*currency* Italian lira
*population* (1990 est) 23,000; growth rate 0.1% p.a.
*life expectancy* men 70, women 77
*language* Italian
*religion* Roman Catholic 95%
*literacy* 97% (1987)
*chronology*
*1862* Independence recognized under Italy's protection.
*1947–86* Governed by a series of left-wing and centre-left coalitions.
*1986* Communist and Christian Democrat 'grand coalition'.
*1992* Joined the United Nations.

# São Tomé e Príncipe
Democratic
Republic of

*area* 1,000 sq km/386 sq mi
*capital* São Tomé
*towns* Santo Antonio, Santa Cruz
*physical* comprises two main islands and several smaller ones, all volcanic; thickly forested and fertile
*head of state and government* Miguel Trovoada from 1991
*political system* emergent democratic republic
*exports* cocoa, copra, coffee, palm oil and kernels
*currency* dobra
*population* (1990 est) 125,000; growth rate 2.5% p.a.
*life expectancy* men 62, women 62
*languages* Portuguese (official), Fang (Bantu)
*religions* Roman Catholic 80%, animist
*literacy* men 73%, women 42% (1981)
*GNP* $32 million (1987); $384 per head (1986)
*chronology*
*1471* Discovered by Portuguese.
*1522–1973* A province of Portugal.
*1973* Achieved internal self-government.
*1975* Independence achieved from Portugal, with Manuel Pinto da Costa as president.

*1984* Formally declared a nonaligned state.
*1987* Constitution amended.
*1988* Unsuccessful coup attempt against da Costa.
*1990* New constitution approved.
*1991* First multiparty elections held; Miguel Trovoada replaced Pinto da Costa.

## Saudi Arabia
Kingdom of
(*al-Mamlaka al-'Arabiya as-Sa'udiya*)

***area*** 2,200,518 sq km/849,400 sq mi
***capital*** Riyadh
***towns*** Mecca, Medina, Taif, Jidda, Dammam
***physical*** desert, sloping to the Persian Gulf from a height of 2,750 m/9,000 ft in the W
***environment*** oil pollution caused by the Gulf War 1990–91 has affected 460 km/285 mi of the Saudi coastline, threatening desalination plants and damaging the wildlife of saltmarshes, mangrove forest, and mudflats
***head of state and government*** King Fahd Ibn Abdul Aziz from 1982
***political system*** absolute monarchy
***exports*** oil, petroleum products
***currency*** rial
***population*** (1990 est) 16,758,000 (16% nomadic); growth rate 3.1% p.a.
***life expectancy*** men 64, women 67 (1989)
***language*** Arabic
***religion*** Sunni Muslim; there is a Shi'ite minority
***literacy*** men 34%, women 12% (1980 est)
***GNP*** $70 bn (1988); $6,170 per head (1988)

*chronology*
*1926–32* Territories of Nejd and Hejaz united and kingdom established.
*1953* King Ibn Saud died and was succeeded by his eldest son, Saud.
*1964* King Saud forced to abdicate; succeeded by his brother, Faisal.
*1975* King Faisal assassinated; succeeded by his half-brother, Khalid.
*1982* King Khalid died; succeeded by his brother, Crown Prince Fahd.
*1987* Rioting by Iranian pilgrims caused 400 deaths in Mecca; diplomatic relations with Iran severed.
*1990* Iraqi troops invaded and annexed Kuwait and massed on Saudi Arabian border. King Fahd called for help from US and UK forces.
*1991* King Fahd provided military and financial assistance in Gulf War. Calls from religious leaders for 'consultative assembly' to assist in government of kingdom. Saudi Arabia attended Middle East peace conference.
*1992* Formation of a 'consultative council' seen as possible move towards representative government.

## Senegal
Republic of
(*République du Sénégal*)

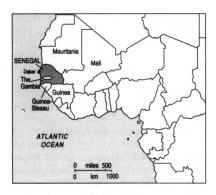

*area* 196,200 sq km/75,753 sq mi
*capital* (and chief port) Dakar
*towns* Thiès, Kaolack
*physical* plains rising to hills in SE; swamp and tropical forest in SW
*head of state and government* Abdou Diouf from 1981
*political system* emergent socialist democratic republic
*exports* peanuts, cotton, fish, phosphates
*currency* franc CFA
*population* (1990 est) 7,740,000; growth rate 3.1% p.a.
*life expectancy* men 51, women 54 (1989)
*languages* French (official); African dialects are spoken
*religions* Muslim 80%, Roman Catholic 10%, animist
*literacy* men 37%, women 19% (1985 est)
*GNP* $2 bn (1987); $380 per head (1984)
*chronology*
*1659* Became a French colony.
*1854–65* Interior occupied by French.
*1902* Became a territory of French West Africa.
*1959* Formed the Federation of Mali with French Sudan.
*1960* Independence achieved from France, but withdrew from the federation. Léopold Sédar Senghor, leader of the Senegalese

Progressive Union (UPS), became president.
*1966* UPS declared the only legal party.
*1974* Pluralist system re-established.
*1976* UPS reconstituted as Senegalese Socialist Party (PS). Prime Minister Abdou Diouf nominated as Senghor's successor.
*1980* Senghor succeeded by Diouf. Troops sent to defend Gambia.
*1981* Military help again sent to Gambia.
*1982* Confederation of Senegambia came into effect.
*1983* Diouf re-elected. Post of prime minister abolished.
*1988* Diouf decisively re-elected.
*1989* Violent clashes between Senegalese and Mauritanians in Dakar and Nouakchott killed more than 450 people; over 50,000 people repatriated from both countries. Senegambia federation abandoned.
*1991* Constitutional changes outlined.
*1992* Diplomatic links with Mauritania re-established.

# Seychelles
Republic of

*area* 453 sq km/175 sq mi
*capital* Victoria (on Mahé island)
*towns* Cascade, Port Glaud, Misere
*physical* comprises two distinct island groups, one concentrated, the other widely scattered, totalling over 100 islands and islets
*head of state and government* France-Albert René from 1977
*political system* one-party socialist republic
*exports* copra, cinnamon
*currency* Seychelles rupee
*population* (1990) 71,000; growth rate 2.2% p.a.
*life expectancy* 66 years (1988)
*languages* creole (Asian, African, European mixture) 95%, English, French (all official)
*religion* Roman Catholic 90%
*literacy* 80% (1989)
*GNP* $175 million; $2,600 per head (1987)
*chronology*
*1744* Became a French colony.
*1794* Captured by British.
*1814* Ceded by France to Britain; incorporated as a dependency of Mauritius.

*1903* Became a separate British colony.
*1975* Internal self-government agreed.
*1976* Independence achieved from Britain as a republic within the Commonwealth, with James Mancham as president.
*1977* France-Albert René ousted Mancham in an armed coup and took over presidency.
*1979* New constitution adopted; Seychelles People's Progressive Front (SPPF) sole legal party.
*1981* Attempted coup by South African mercenaries thwarted.
*1984* René re-elected.
*1987* Coup attempt foiled.
*1989* René re-elected.
*1991* Multiparty politics promised.
*1992* Mancham returned from exile. Constitutional commission elected; referendum on constitutional reform received insufficient support.

**Sierra Leone**
Republic of

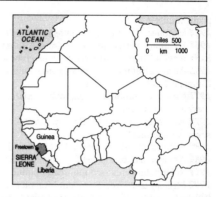

*area* 71,740 sq km/27,710 sq mi
*capital* Freetown
*towns* Koidu, Bo, Kenema, Makeni
*physical* mountains in E; hills and forest; coastal mangrove swamps
*head of state and government* military council headed by Capt Valentine Strasser from 1992
*political system* transitional
*exports* palm kernels, cocoa, coffee, ginger, diamonds, bauxite, rutile
*currency* leone
*population* (1990 est) 4,168,000; growth rate 2.5% p.a.
*life expectancy* men 41, women 47 (1989)
*languages* English (official), local languages
*religions* animist 52%, Muslim 39%, Protestant 6%, Roman Catholic 2% (1980 est)
*literacy* men 38%, women 21% (1985 est)
*GNP* $965 million (1987); $320 per head (1984)
*chronology*
*1808* Became a British colony.
*1896* Hinterland declared a British protectorate.
*1961* Independence achieved from Britain within the Commonwealth, with Milton Margai, leader of Sierra Leone People's Party (SLPP), as

prime minister.
*1964* Milton succeeded by his half-brother, Albert Margai.
*1967* Election results disputed by army, who set up a National Reformation Council and forced the governor general to leave.
*1968* Army revolt made Siaka Stevens, leader of the All People's Congress (APC), prime minister.
*1971* New constitution adopted, making Sierra Leone a republic, with Stevens as president.
*1978* APC declared only legal party. Stevens sworn in for another seven-year term.
*1985* Stevens retired; succeeded by Maj-Gen Joseph Momoh.
*1989* Attempted coup against President Momoh foiled.
*1991* Referendum endorsed multiparty politics.
*1992* Military take-over; President Momoh fled. National Provisional Ruling Council (NPRC) established under Capt Valentine Strasser.

## Singapore
Republic of

***area*** 622 sq km/240 sq mi
***capital*** Singapore City
***towns*** Jurong, Changi
***physical*** comprises Singapore Island, low and flat, and 57 small islands
***head of state*** Wee Kim Wee from 1985
***head of government*** Goh Chok Tong from 1990
***political system*** liberal democracy with strict limits on dissent
***exports*** electronics, petroleum products, rubber, machinery, vehicles
***currency*** Singapore dollar
***population*** (1990 est) 2,703,000 (Chinese 75%, Malay 14%, Tamil 7%); growth rate 1.2% p.a.
***life expectancy*** men 71, women 77 (1989)
***languages*** Malay (national tongue), Chinese, Tamil, English (all official)
***religions*** Buddhist, Taoist, Muslim, Hindu, Christian
***literacy*** men 93%, women 79% (1985 est)

***GDP*** $19.9 bn (1987); $7,616 per head

***chronology***

*1819* Singapore leased to British East India Company.
*1858* Placed under rule of British crown.
*1942* Invaded and occupied by Japan.
*1945* Japanese removed by British forces.
*1959* Independence achieved from Britain; Lee Kuan Yew became prime minister.
*1963* Joined new Federation of Malaysia.
*1965* Left federation to become an independent republic.
*1984* Opposition made advances in parliamentary elections.
*1986* Opposition leader convicted of perjury and prohibited from standing for election.
*1988* Ruling conservative party elected to all but one of available assembly seats; increasingly authoritarian rule.
*1990* Lee Kuan Yew resigned as prime minister; replaced by Goh Chok Tong.
*1991* People's Action Party (PAP) and Goh Chok Tong re-elected.

# Slovak Republic
(*Slovenská Republika*)

***area*** 49,035 sq km/18,940 sq mi
***capital*** Bratislava
***towns*** Kosice, Nitra, Presov, Banská Bystrica
***physical*** Carpathian Mountains including Tatra and Beskids in N; fine beech and oak forests; Danube plain in S
***head of state*** Michal Kovak from 1993
***head of government*** Vladimir Meciar from 1993
***political system*** emergent democracy
***exports*** iron ore, copper, mercury, magnesite, armaments, chemicals, textiles, machinery
***currency*** new currency based on koruna
***population*** (1991) 5,268,900 (with Hungarian and other minorities); growth rate 0.4% p.a.
***life expectancy*** men 68, women 75
***languages*** Slovak (official)
***religions*** Roman Catholic (over 50%), Lutheran, Reformist, Orthodox
***literacy*** 100%
***GDP*** $10,000 million (1990); $1,887 per head

*chronology*
**906–1918** Under Magyar domination.
**1918** Independence achieved from Austro-Hungarian Empire; Slovaks joined Czechs in forming Czechoslovakia as independent nation.
**1948** Communists assumed power in Czechoslovakia.
**1968** Slovak Socialist Republic created under new federal constitution.
**1989** Pro-democracy demonstrations in Prague and Bratislava; new political parties formed, including Slovak-based People Against Violence (PAV); Communist Party stripped of powers. Dec: new 'grand coalition' government formed, including former dissidents; political parties legalized; Václav Havel appointed state president. Amnesty granted to 22,000 prisoners; calls for USSR to withdraw its troops.
**1990** July: Havel re-elected president in multiparty elections.
**1991** Evidence of increasing Czech and Slovak separatism. March: PAV splinter group formed under Slovak premier Vladimir Meciar. April: Meciar dismissed, replaced by Jan Carnogursky; pro-Meciar rallies held in Bratislava. July: Soviet troops withdrawn. Oct: Public Against Violence renamed Civic Democratic Union–Public Against Violence (PAV).
**1992** March: PAV renamed Civic Democratic Union (CDU).
June: Havel resigned following Slovak gains in assembly elections.
Aug: agreement on creation of separate Czech and Slovak states from Jan 1993.
**1993** Jan: Slovak Republic became sovereign state, with Meciar, leader of the Movement for a Democratic Slovakia (MFDS), as prime minister. Feb: Michal Kovak became president.

# Slovenia
Republic of

*area* 20,251 sq km/7,817 sq mi
*capital* Ljubljana
*towns* Maribor, Kranj, Celji, Koper
*physical* mountainous; Sava and Drava rivers
*head of state* Milan Kucan from 1990
*head of government* Janez Drnovsek from 1992
*political system* emergent democracy
*products* grain, sugarbeet, livestock, timber, cotton and woollen textiles, steel, vehicles
*currency* tolar
*population* (1990) 2,000,000 (Slovene 91%, Croat 3%, Serb 2%)
*languages* Slovene, resembling Serbo-Croat
*religion* Roman Catholic
*chronology*
*1918* United with Serbia and Croatia.
*1929* The kingdom of Serbs, Croats, and Slovenes took the name of Yugoslavia.
*1945* Became a constituent republic of Yugoslav Socialist Federal Republic.

***mid-1980s*** The Slovenian Communist Party liberalized itself and agreed to free elections. Yugoslav counterintelligence (KOV) began repression.
***1989*** Jan: Social Democratic Alliance of Slovenia launched as the first political organization independent of Communist Party.
Sept: constitution changed to allow secession from federation.
***1990*** Feb: Slovene League of Communists, renamed as the Party of Democratic Reform, severed its links with the Yugoslav League of Communists. April: nationalist Democratic Opposition of Slovenia (DEMOS) coalition secured victory in first multiparty parliamentary elections; Milan Kucan became president. July: sovereignty declared. Dec: independence overwhelmingly approved in referendum.
***1991*** June: independence declared; 100 killed after federal army intervened; cease-fire brokered by European Community (EC). July: cease-fire agreed between federal troops and nationalists. Oct: withdrawal of Yugoslav army completed. Dec: DEMOS coalition dissolved.
***1992*** Jan: EC recognized Slovenia's independence. April: Janez Drnovsek appointed prime minister designate; independence recognized by USA. May: admitted into United Nations and Conference on Security and Cooperation in Europe (CSCE). Dec: Liberal Democrats and Christian Democrats won assembly elections; Kucan re-elected president.
***1993*** Drnovsek re-elected prime minister.

# Solomon Islands

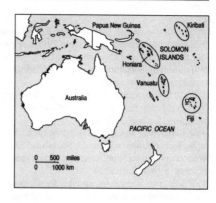

*area* 27,600 sq km/10,656 sq mi
*capital* Honiara (on Guadalcanal)
*towns* Gizo, Yandina
*physical* comprises all but the northernmost islands (which belong to Papua New Guinea) of a Melanesian archipelago stretching nearly 1,500 km/900 mi. The largest is Guadalcanal (area 6,500 sq km/2,510 sq mi); others are Malaita, San Cristobal, New Georgia, Santa Isabel, Choiseul; mainly mountainous and forested
*head of state* Elizabeth II represented by governor general
*head of government* Solomon Mamaloni from 1989
*political system* constitutional monarchy
*exports* fish products, palm oil, copra, cocoa, timber
*currency* Solomon Island dollar
*population* (1990 est) 314,000 (Melanesian 95%, Polynesian 4%); growth rate 3.9% p.a.
*life expectancy* men 66, women 71
*languages* English (official); there are some 120 Melanesian dialects
*religions* Anglican 34%, Roman Catholic 19%, South Sea Evangelical 17%
*literacy* 60% (1989)
*GNP* $141 million; $420 per head (1987)

## chronology

**1893** Solomon Islands placed under British protection.
**1978** Independence achieved from Britain within the Commonwealth, with Peter Kenilorea as prime minister.
**1981** Solomon Mamaloni of the People's Progressive Party replaced Kenilorea as prime minister.
**1984** Kenilorea returned to power, heading a coalition government.
**1986** Kenilorea resigned after allegations of corruption; replaced by his deputy, Ezekiel Alebua.
**1988** Kenilorea elected deputy prime minister. Joined Vanuatu and Papua New Guinea to form the Spearhead Group, aiming to preserve Melanesian cultural traditions and secure independence for the French territory of New Caledonia.
**1989** Solomon Mamaloni, now leader of the People's Alliance Party (PAP), elected prime minister; formed PAP-dominated coalition.
**1990** Mamaloni resigned as PAP party leader, but continued as head of a government of national unity.

# Somalia
Somali Democratic Republic
(*Jamhuriyadda Dimugradiga Somaliya*)

*area* 637,700 sq km/246,220 sq mi
*capital* Mogadishu
*towns* Hargeisa, Kismayu, Berbera
*physical* mainly flat, with hills in N
*environment* destruction of trees for fuel and by grazing livestock has led to an increase in desert area
*head of state and government* Ali Mahdi Mohammed from 1991
*political system* one-party socialist republic
*exports* livestock, skins, hides, bananas, fruit
*currency* Somali shilling
*population* (1990 est) 8,415,000 (including 350,000 refugees in Ethiopia and 50,000 in Djibouti); growth rate 3.1% p.a.
*life expectancy* men 53, women 53 (1989)
*languages* Somali, Arabic (both official), Italian, English
*religion* Sunni Muslim 99%
*literacy* 40% (1986)
*GNP* $1.5 bn; $290 per head (1987)
*chronology*
*1884–87* British protectorate of Somaliland established.
*1889* Italian protectorate of Somalia established.
*1960* Independence achieved from Italy and Britain.

**1963** Border dispute with Kenya; diplomatic relations broken with Britain.
**1968** Diplomatic relations with Britain restored.
**1969** Army coup led by Maj-Gen Mohamed Siad Barre; constitution suspended, Supreme Revolutionary Council set up; name changed to Somali Democratic Republic.
**1978** Defeated in eight-month war with Ethiopia. Armed insurrection began in north.
**1979** New constitution for socialist one-party state adopted.
**1982** Antigovernment Somali National Movement formed. Oppressive countermeasures by government.
**1987** Barre re-elected president.
**1989** Dissatisfaction with government and increased guerrilla activity in north.
**1990** Civil war intensified. Constitutional reforms promised.
**1991** Mogadishu captured by rebels; Barre fled; Ali Mahdi Mohammed named president; free elections promised. Secession of NE Somalia, as the Somaliland Republic, announced. Cease-fire signed, but later collapsed. Thousands of casualties as a result of heavy fighting in capital.
**1992** Relief efforts to ward off impending famine severely hindered by unstable political situation; relief convoys hijacked by 'war lords'. Dec: UN peacekeeping troops, mainly US Marines, drafted in to protect relief operations; dominant warlords agreed truce.
**1993** March: leaders of armed factions agreed to federal system of government, based on 18 autonomous regions.

# South Africa
Republic of
(*Republiek van Suid-Afrika*)

***area*** 1,223,181 sq km/472,148 sq mi (includes Walvis Bay and independent black homelands)
***capital*** Cape Town (legislative), Pretoria (administrative), Bloemfontein (judicial)
***towns*** Johannesburg, Durban, Port Elizabeth, East London
***physical*** southern end of large plateau, fringed by mountains and lowland coastal margin
***territories*** Marion Island and Prince Edward Island in the Antarctic
***head of state and government*** F W de Klerk from 1989
***political system*** nationalist republic, restricted democracy
***exports*** maize, sugar, fruit, wool, gold (world's largest producer), platinum, diamonds, uranium, iron and steel, copper; mining and minerals are largest export industry, followed by arms manufacturing
***currency*** rand
***population*** (1990 est) 39,550,000 (73% black: Zulu, Xhosa, Sotho, Tswana; 18% white: 3% mixed, 3% Asian); growth rate 2.5% p.a.
***life expectancy*** whites 71, Asians 67, blacks 58
***languages*** Afrikaans and English (both official), Bantu

*religions* Dutch Reformed Church 40%, Anglican 11%, Roman Catholic 8%, other Christian 25%, Hindu, Muslim
*literacy* whites 99%, Asians 69%, blacks 50% (1989)
*GNP* $81 bn; $1,890 per head (1987)
*chronology*
*1910* Union of South Africa formed from two British colonies and two Boer republics.
*1912* African National Congress (ANC) formed.
*1948* Apartheid system of racial discrimination initiated by Daniel Malan, leader of National Party (NP).
*1955* Freedom Charter adopted by ANC.
*1958* Malan succeeded as prime minister by Hendrik Verwoerd.
*1960* ANC banned.
*1961* South Africa withdrew from Commonwealth and became a republic.
*1962* ANC leader Nelson Mandela jailed.
*1964* Mandela, Walter Sisulu, Govan Mbeki, and five other ANC leaders sentenced to life imprisonment.
*1966* Verwoerd assassinated; succeeded by B J Vorster.
*1976* Soweto uprising.
*1977* Death in custody of Pan African Congress (PAC) activist Steve Biko.
*1978* Vorster resigned and was replaced by P W Botha.
*1984* New constitution adopted, giving segregated representation to Coloureds and Asians and making Botha president. Nonaggression pact with Mozambique signed but not observed.
*1985* Growth of violence in black townships.
*1986* Commonwealth agreed on limited sanctions. US Congress voted to impose sanctions. Some major multinational companies closed down their South African operations.
*1987* Government formally acknowledged the presence of its military forces in Angola.
*1988* Botha announced 'limited constitutional reforms'. South Africa agreed to withdraw from Angola and recognize Namibia's independence as part of regional peace accord.
*1989* Botha gave up NP leadership and state presidency. F W de Klerk

became president. ANC activists released; beaches and public facilities desegregated. Elections held in Namibia to create independence government.
***1990*** ANC ban lifted; Nelson Mandela released from prison. NP membership opened to all races. ANC leader Oliver Tambo returned. Daily average of 35 murders and homicides recorded.
***1991*** Mandela and Zulu leader Chief Gatsha Buthelezi urged end to fighting between ANC and Inkatha. Mandela elected ANC president. Revelations of government support for Inkatha threatened ANC cooperation. De Klerk announced repeal of remaining apartheid laws. South Africa readmitted to international sport. USA lifted sanctions. PAC and Buthelezi withdrew from negotiations over new constitution.
***1992*** Constitution leading to all-races majority rule approved by whites-only referendum. Massacre of civilians at black township of Boipathong near Johannesburg by Inkatha, aided and abetted by police, threatened constitutional talks.
***1993*** Feb: de Klerk and Nelson Mandela agreed to formation of government of national unity after free elections late 1993/early 1994. Buthelezi not consulted; he opposed such an arrangement. April: ANC leader Chris Hani assassinated; Andries Treurnicht, leader of Conservative Party, and Oliver Tambo died.

## Spain
(*España*)

*area* 504,750 sq km/194,960 sq mi
*capital* Madrid
*towns* Seville, Barcelona, Valencia, Málaga, Cádiz
*physical* central plateau with mountain ranges; lowlands in S
*territories* Balearic and Canary Islands; in N Africa: Ceuta, Melilla, Alhucemas, Chafarinas Is, Peñón de Vélez
*head of state* King Juan Carlos I from 1975
*head of government* Felipe González Márquez from 1982
*political system* constitutional monarchy
*exports* citrus fruits, grapes, pomegranates, vegetables, wine, sherry, olive oil, canned fruit and fish, iron ore, cork, vehicles, textiles, petroleum products, leather goods, ceramics
*currency* peseta
*population* (1990 est) 39,623,000; growth rate 0.2% p.a.
*life expectancy* men 74, women 80 (1989)
*languages* Spanish (Castilian, official), Basque, Catalan, Galician, Valencian, Majorcan
*religion* Roman Catholic 99%
*literacy* 97% (1989)

*GNP* $288 bn (1987); $4,490 per head (1984)
*recent chronology*
*1936–39* Civil war; General Francisco Franco became head of state and government; fascist party Falange declared only legal political organization.
*1947* General Franco announced restoration of the monarchy after his death, with Prince Juan Carlos as his successor.
*1975* Franco died; succeeded as head of state by King Juan Carlos I.
*1978* New constitution adopted with Adolfo Suárez, leader of the Democratic Centre Party, as prime minister.
*1981* Suárez resigned; succeeded by Leopoldo Calvo Sotelo. Attempted military coup thwarted.
*1982* Socialist Workers' Party (PSOE), led by Felipe González, won a sweeping electoral victory. Basque separatist organization (ETA) stepped up its guerrilla campaign.
*1985* ETA's campaign spread to holiday resorts.
*1986* Referendum confirmed NATO membership. Spain joined the European Economic Community.
*1988* Spain joined the Western European Union.
*1989* PSOE lost seats to hold only parity after general election. Talks between government and ETA collapsed and truce ended.
*1992* ETA's 'armed struggle' resumed. Nov: Maastricht Treaty ratified by parliament.

## Sri Lanka
Democratic Socialist Republic of (*Prajathanrika Samajawadi Janarajaya Sri Lanka*)
(until 1972 *Ceylon*)

*area* 65,600 sq km/25,328 sq mi
*capital* (and chief port) Colombo
*towns* Kandy, Jaffna, Galle, Negombo, Trincomalee
*physical* flat in N and around the coast; hills and mountains in S and central interior
*head of state* Ranasinghe Premadasa from 1989
*head of government* Dingiri Banda Wijetunge from 1989
*political system* liberal democratic republic
*exports* tea, rubber, coconut products, graphite, sapphires, rubies, other gemstones
*currency* Sri Lanka rupee
*population* (1990 est) 17,135,000 (Sinhalese 74%, Tamils 17%, Moors 7%); growth rate 1.8% p.a.
*life expectancy* men 67, women 72 (1989)
*languages* Sinhala, Tamil, English
*religions* Buddhist 69%, Hindu 15%, Muslim 8%, Christian 7%
*literacy* 87% (1988)
*GNP* $7.2 bn; $400 per head (1988)

*chronology*
*1802* Ceylon became a British colony.
*1948* Ceylon achieved independence from Britain within the Commonwealth.
*1956* Sinhala established as the official language.
*1959* Prime Minister Solomon Bandaranaike assassinated.
*1972* Socialist Republic of Sri Lanka proclaimed.
*1978* Presidential constitution adopted by new government led by Junius Jayawardene of the United National Party (UNP).
*1983* Tamil guerrilla violence escalated; state of emergency imposed.
*1987* President Jayawardene and Indian prime minister Rajiv Gandhi signed Colombo Accord. Violence continued despite cease-fire policed by Indian troops.
*1988* Left-wing guerrillas campaigned against Indo-Sri Lankan peace pact. Prime Minister Ranasinghe Premadasa elected president.
*1989* Premadasa became president; D B Wijetunge, prime minister. Leaders of the Tamil United Liberation Front (TULF) and terrorist People's Liberation Front assassinated.
*1990* Indian peacekeeping force withdrawn. Violence continued.
*1991* March: defence minister, Ranjan Wijeratne, assassinated; Sri Lankan army killed 2,552 Tamil Tigers at Elephant Pass. October: impeachment motion against President Premadasa failed. Dec: new party, the Democratic United National Front, formed by former members of UNP.

# Sudan

**Sudan**
Democratic
Republic of
(*Jamhuryat
es-Sudan*)

*area* 2,505,800 sq km/967,489 sq mi
*capital* Khartoum
*towns* Omdurman, Juba, Wadi Medani, Kassala, Atbara, Kosti, Port Sudan
*physical* fertile valley of river Nile separates Libyan Desert in W from high rocky Nubian Desert in E
*environment* the building of the Jonglei Canal to supply water to N Sudan and Egypt threatens the grasslands of S Sudan
*head of state and government* General Omar Hassan Ahmed el-Bashir from 1989
*political system* military republic
*exports* cotton, gum arabic, sesame seed, peanuts, sorghum
*currency* Sudanese pound
*population* (1990 est) 25,164,000; growth rate 2.9% p.a.
*life expectancy* men 51, women 55 (1989)
*languages* Arabic 51% (official), local languages
*religions* Sunni Muslim 73%, animist 18%, Christian 9% (in south)
*literacy* 30% (1986)
*GNP* $8.5 bn (1988); $330 per head (1988)
*recent chronology*
*1956* Sudan achieved independence from Britain and Egypt as

a republic.
*1958* Military coup replaced civilian government with Supreme Council of the Armed Forces.
*1964* Civilian rule reinstated.
*1969* Coup led by Col Gaafar Mohammed Nimeri established Revolutionary Command Council (RCC); name changed to Democratic Republic of Sudan.
*1970* Union with Egypt agreed in principle.
*1971* New constitution adopted; Nimeri confirmed as president; Sudanese Socialist Union (SSU) declared only legal party.
*1972* Proposed Federation of Arab Republics, comprising Sudan, Egypt, and Syria, abandoned. Addis Ababa conference proposed autonomy for southern provinces.
*1974* National assembly established.
*1983* Nimeri re-elected. Shari'a (Islamic law) introduced.
*1985* Nimeri deposed in a bloodless coup led by General Swar al-Dahab; transitional military council set up. State of emergency declared.
*1986* More than 40 political parties fought general election; coalition government formed.
*1987* Virtual civil war with Sudan People's Liberation Army (SPLA).
*1988* Al-Mahdi formed a new coalition. Another flare-up of civil war between north and south created tens of thousands of refugees. Floods made 1.5 million people homeless. Peace pact signed with SPLA.
*1989* Sadiq al-Mahdi overthrown in coup led by General Omar Hassan Ahmed el-Bashir.
*1990* Civil war continued with new SPLA offensive.
*1991* Federal system introduced, with division of country into nine states.
*1993* March: SPLA leaders John Garang and Riek Machar announced unilateral cease-fire in ten years' war with government in Khartoum.

# Surinam
Republic of
(*Republiek Suriname*)

*area* 163,820 sq km/63,243 sq mi
*capital* Paramaribo
*towns* Nieuw Nickerie, Brokopondo, Nieuw Amsterdam
*physical* hilly and forested, with flat and narrow coastal plain
*head of state and government* Ronald Venetiaan from 1991
*political system* emergent democratic republic
*exports* alumina, aluminium, bauxite, rice, timber
*currency* Surinam guilder
*population* (1990 est) 408,000 (Hindu 37%, Creole 31%, Javanese 15%); growth rate 1.1% p.a.
*life expectancy* men 66, women 71 (1989)
*languages* Dutch (official), Sranan (creole), English, others
*religions* Christian 30%, Hindu 27%, Muslim 20%
*literacy* 65% (1989)
*GNP* $1.1 bn (1987); $2,920 per head (1985)
*chronology*
*1667* Became a Dutch colony.
*1954* Achieved internal self-government as Dutch Guiana.
*1975* Independence achieved from the Netherlands, with Dr Johan

Ferrier as president and Henck Arron as prime minister; 40% of the population emigrated to the Netherlands.
*1980* Arron's government overthrown in army coup; Ferrier refused to recognize military regime; appointed Dr Henk Chin A Sen to lead civilian administration. Army replaced Ferrier with Dr Chin A Sen.
*1982* Army, led by Lt Col Desi Bouterse, seized power, setting up a Revolutionary People's Front.
*1985* Ban on political activities lifted.
*1986* Antigovernment rebels brought economic chaos to Surinam.
*1987* New constitution approved.
*1988* Ramsewak Shankar elected president.
*1989* Bouterse rejected peace accord reached by President Shankar with guerrilla insurgents, vowed to continue fighting.
*1990* Shankar deposed in army coup.
*1991* Johan Kraag became interim president. New Front for Democracy won assembly majority. Ronald Venetiaan elected president.
*1992* Peace accord with guerrilla groups.

# Swaziland

**Swaziland**
Kingdom of
(*Umbuso weSwatini*)

*area* 17,400 sq km/6,716 sq mi
*capital* Mbabane
*towns* Manzini, Big Bend
*physical* central valley; mountains in W (Highveld); plateau in E (Lowveld and Lubombo plateau)
*head of state and government* King Mswati III from 1986
*political system* near-absolute monarchy
*exports* sugar, canned fruit, wood pulp, asbestos
*currency* lilangeni
*population* (1990 est) 779,000; growth rate 3% p.a.
*life expectancy* men 47, women 54 (1989)
*languages* Swazi 90%, English (both official)
*religions* Christian 57%, animist
*literacy* men 70%, women 66% (1985 est)
*GNP* $539 million; $750 per head (1987)
*chronology*
*1903* Swaziland became a special High Commission territory.

*1967* Achieved internal self-government.
*1968* Independence achieved from Britain, within the Commonwealth, as the Kingdom of Swaziland, with King Sobhuza II as head of state.
*1973* The king suspended the constitution and assumed absolute powers.
*1978* New constitution adopted.
*1982* King Sobhuza died; his place was taken by one of his wives, Dzeliwe, until his son, Prince Makhosetive, reached the age of 21.
*1983* Queen Dzeliwe ousted by another wife, Ntombi.
*1984* After royal power struggle, it was announced that the crown prince would become king at 18.
*1986* Crown prince formally invested as King Mswati III.
*1987* Power struggle developed between advisory council Liqoqo and Queen Ntombi over accession of king. Mswati dissolved parliament; new government elected with Sotsha Dlamini as prime minister.
*1991* Calls for democratic reform.
*1992* Mswati dissolved parliament, assuming 'executive powers'.

## Sweden
Kingdom of
(*Konungariket Sverige*)

*area* 450,000 sq km/173,745 sq mi
*capital* Stockholm
*towns* Göteborg, Malmö, Uppsala, Norrköping, Västerås
*physical* mountains in W; plains in S; thickly forested; more than 20,000 islands off the Stockholm coast
*environment* of the country's 90,000 lakes, 20,000 are affected by acid rain; 4,000 are so severely acidified that no fish are thought to survive in them
*head of state* King Carl XVI Gustaf from 1973
*head of government* Carl Bildt from 1991
*political system* constitutional monarchy
*exports* aircraft, vehicles, missiles, electronics, petrochemicals, textiles, furnishings, ornamental glass, paper, iron and steel
*currency* krona
*population* (1990 est) 8,407,000 (including 17,000 Saami (Lapps) and 1.2 million postwar immigrants from Finland, Turkey, Yugoslavia, Greece, Iran, other Nordic countries); growth rate 0.1% p.a.
*life expectancy* men 74, women 81 (1989)
*languages* Swedish; there are Finnish- and Saami-speaking minorities
*religion* Lutheran (official) 95%
*literacy* 99% (1989)

***GNP*** $179 bn; $11,783 per head (1989)
***chronology***
***12th century*** United as an independent nation.
***1397–1520*** Under Danish rule.
***1914–45*** Neutral in both world wars.
***1951–76*** Social Democratic Labour Party (SAP) in power.
***1969*** Olof Palme became SAP leader and prime minister.
***1971*** Constitution amended, creating a single-chamber Riksdag, the governing body.
***1975*** Monarch's last constitutional powers removed.
***1976*** Thorbjörn Fälldin, leader of the Centre Party, became prime minister, heading centre-right coalition.
***1982*** SAP, led by Palme, returned to power.
***1985*** SAP formed minority government, with Communist support.
***1986*** Olof Palme murdered. Ingvar Carlsson became prime minister and SAP party leader.
***1988*** SAP re-elected with reduced majority; Green Party gained representation in Riksdag.
***1990*** SAP government resigned. Sweden to apply for European Community (EC) membership.
***1991*** Formal application for EC membership submitted. Election defeat for SAP; Carlsson resigned. Coalition government formed; Carl Bildt became new prime minister.
***1992*** Cross-party agreement to solve economic problems.

## Switzerland

Swiss Confederation
(German *Schweiz*,
French *Suisse*,
Romansch *Svizzera*)

*area* 41,300 sq km/15,946 sq mi
*capital* Bern
*towns* Zürich, Geneva, Lausanne, Basel
*physical* most mountainous country in Europe (Alps and Jura mountains); highest peak Dufourspitze 4,634 m/15,203 ft
*environment* an estimated 43% of coniferous trees, particularly in the central alpine region, have been killed by acid rain, 90% of which comes from other countries. Over 50% of bird species are classified as threatened
*head of state and government* Adolf Ogi from 1993
*government* federal democratic republic
*exports* electrical goods, chemicals, pharmaceuticals, watches, precision instruments, confectionery
*currency* Swiss franc
*population* (1990 est) 6,628,000; growth rate 0.2% p.a.
*life expectancy* men 74, women 82 (1989)
*languages* German 65%, French 18%, Italian 12%, Romansch 1% (all official)

*religions* Roman Catholic 50%, Protestant 48%
*literacy* 99% (1989)
*GNP* $111 bn (1988); $26,309 per head (1987)
*chronology*
*1648* Became independent of the Holy Roman Empire.
*1798–1815* Helvetic Republic established by French revolutionary armies.
*1847* Civil war resulted in greater centralization.
*1874* Principle of the referendum introduced.
*1971* Women given the vote in federal elections.
*1984* First female cabinet minister appointed.
*1986* Referendum rejected proposal for membership of United Nations.
*1989* Referendum supported abolition of citizen army and military service requirements.
*1991* 18-year-olds allowed to vote for first time in national elections. Four-party coalition remained in power.
*1992* René Felber elected president with Adolf Ogi as vice president. Closer ties with European Community (EC) rejected in national referendum.
*1993* Ogi replaced Felber as head of state.

# Syria
Syrian Arab
Republic
(*al-Jamhuriya al-Arabya as-Suriya*)

*area* 185,200 sq km/71,506 sq mi
*capital* Damascus
*towns* Aleppo, Homs, Hama, Latakia
*physical* mountains alternate with fertile plains and desert areas; Euphrates River
*head of state and government* Hafez al-Assad from 1971
*political system* socialist republic
*exports* cotton, cereals, oil, phosphates, tobacco
*currency* Syrian pound
*population* (1990 est) 12,471,000; growth rate 3.5% p.a.
*life expectancy* men 67, women 69 (1989)
*languages* Arabic 89% (official), Kurdish 6%, Armenian 3%
*religions* Sunni Muslim 74%; ruling minority Alawite, and other Islamic sects 16%; Christian 10%
*literacy* men 76%, women 43% (1985 est)
*GNP* $17 bn (1986); $702 per head
*chronology*
*1946* Achieved full independence from France.
*1958* Merged with Egypt to form the United Arab Republic (UAR).

*1961* UAR disintegrated.
*1967* Six-Day War resulted in the loss of territory to Israel.
*1970–71* Syria supported Palestinian guerrillas against Jordanian troops.
*1971* Following a bloodless coup, Hafez al-Assad became president.
*1973* Israel consolidated its control of the Golan Heights after the Yom Kippur War.
*1976* Substantial numbers of troops committed to the civil war in Lebanon.
*1978* Assad re-elected.
*1981–82* Further military engagements in Lebanon.
*1982* Islamic militant uprising suppressed; 5,000 dead.
*1984* Presidents Assad and Gemayel approved plans for government of national unity in Lebanon.
*1985* Assad secured the release of 39 US hostages held in an aircraft hijacked by extremist Shi'ite group, Hezbollah. Assad re-elected.
*1987* Improved relations with USA and attempts to secure the release of Western hostages in Lebanon.
*1989* Diplomatic relations with Morocco restored. Continued fighting in Lebanon; Syrian forces reinforced in Lebanon; diplomatic relations with Egypt restored.
*1990* Diplomatic relations with Britain restored.
*1991* Syria fought against Iraq in Gulf War. President Assad agreed to US Middle East peace plan. Assad re-elected as president.

## Taiwan
Republic of China
(*Chung Hua Min Kuo*)

*area* 36,179 sq km/13,965 sq mi
*capital* Taipei
*towns* Kaohsiung, Keelung
*physical* island (formerly Formosa) off People's Republic of China; mountainous, with lowlands in W
*environment* industrialization has taken its toll: an estimated 30% of the annual rice crop is dangerously contaminated with mercury, cadmium, and other heavy metals
*head of state* Lee Teng-hui from 1988
*head of government* Lien Chan from 1993
*political system* emergent democracy
*exports* textiles, steel, plastics, electronics, foodstuffs
*currency* New Taiwan dollar
*population* (1990) 20,454,000 (Taiwanese 84%, mainlanders 14%); growth rate 1.4% p.a.
*life expectancy* 70 men, 75 women (1986)
*languages* Mandarin Chinese (official); Taiwan, Hakka dialects
*religions* officially atheist; Taoist, Confucian, Buddhist, Christian
*literacy* 90% (1988)
*GNP* $119.1 bn; $6,200 per head (1988)

*chronology*
*1683* Taiwan (Formosa) annexed by China.
*1895* Ceded to Japan.
*1945* Recovered by China.
*1949* Flight of Nationalist government to Taiwan after Chinese communist revolution.
*1954* US-Taiwanese mutual defence treaty.
*1971* Expulsion from United Nations.
*1972* Commencement of legislature elections.
*1975* President Chiang Kai-shek died; replaced as Kuomintang leader by his son, Chiang Ching-kuo.
*1979* USA severed diplomatic relations and annulled 1954 security treaty.
*1986* Democratic Progressive Party (DPP) formed as opposition to the nationalist Kuomintang.
*1987* Martial law lifted; opposition parties legalized; press restrictions lifted.
*1988* President Chiang Ching-kuo died; replaced by Taiwanese-born Lee Teng-hui.
*1989* Kuomintang won assembly elections.
*1990* Formal move towards normalization of relations with China. Hau Pei-tsun became prime minister.
*1991* President Lee Teng-hui declared end to state of civil war with China. Constitution amended. Kuomintang won landslide victory in assembly elections.
*1992* Diplomatic relations with South Korea broken. Dec: in first fully democratic elections Kuomintang lost support to DPP but still secured a majority of seats.
*1993* Lien Chan appointed prime minister.

# Tajikistan

Republic of

*area* 143,100 sq km/55,251 sq mi
*capital* Dushanbe
*towns* Khodzhent (formerly Leninabad), Kurgan-Tyube, Kulyab
*physical* mountainous, more than half of its territory lying above 3,000 m/10,000 ft; huge mountain glaciers, which are the source of many rapid rivers
*head of state* Akbasho Iskandrov from 1992
*head of government* Abdumalik Abdulojonov from 1992
*political system* emergent democracy
*products* fruit, cereals, cotton, cattle , sheep, silks, carpets, coal, lead, zinc, chemicals, oil, gas
*population* (1990) 5,300,000 (Tajik 63%, Uzbek 24%, Russian 8%, Tatar 1%, Kyrgyz 1%, Ukrainian 1%)
*language* Tajik, similar to Farsi (Persian)
*religion* Sunni Muslim
*chronology*
*1921* Part of Turkestan Soviet Socialist Autonomous Republic.

*1929* Became a constituent republic of USSR.
*1990* Ethnic Tajik/Armenian conflict in Dushanbe resulted in rioting against Communist Party of Tajikistan (TCP); state of emergency and curfew imposed.
*1991* Jan: curfew lifted in Dushanbe. March: maintenance of Union endorsed in referendum. Aug: President Makhkamov forced to resign after failed anti-Gorbachev coup; TCP broke links with Moscow. Sept: declared independence; Rakhman Nabiyev elected president; TCP renamed Socialist Party of Tajikistan (SPT); state of emergency declared. Dec: joined new Commonwealth of Independent States (CIS).
*1992* Jan: admitted into Conference for Security and Cooperation in Europe (CSCE). Nabiyev temporarily ousted; state of emergency lifted. Feb: joined the Muslim Economic Cooperation Organization (ECO). March: admitted into United Nations (UN); US diplomatic recognition achieved. May: coalition government formed.
Sept: Nabiyev forced to resign; replaced by Akbasho Iskandrov; Abdumalik Abdulojonov became prime minister.
*1993* Civil war between the forces of the country's communist former rulers and Islamic and pro-democracy groups continued to rage.

## Tanzania

**Tanzania**
United Republic of
(*Jamhuri ya Muungano wa Tanzania*)

*area* 945,000 sq km/364,865 sq mi
*capital* Dodoma (since 1983)
*towns* Zanzibar Town, Mwanza, Dar es Salaam
*physical* central plateau; lakes in N and W; coastal plains; lakes Victoria, Tanganyika, and Niasa
*environment* the black rhino faces extinction as a result of poaching
*head of state and government* Ali Hassan Mwinyi from 1985
*political system* one-party socialist republic
*exports* coffee, cotton, sisal, cloves, tea, tobacco, cashew nuts, diamonds
*currency* Tanzanian shilling
*population* (1990 est) 26,070,000; growth rate 3.5% p.a.
*life expectancy* men 49, women 54 (1989)
*languages* Kiswahili, English (both official)
*religions* Muslim 35%, Christian 35%, traditional 30%
*literacy* 85% (1987)
*GNP* $4.9 bn; $258 per head (1987)

## chronology

**16th–17th centuries** Zanzibar under Portuguese control.
**1890–1963** Zanzibar became a British protectorate.
**1920–46** Tanganyika administered as a British League of Nations mandate.
**1946–62** Tanganyika came under United Nations (UN) trusteeship.
**1961** Tanganyika achieved independence from Britain, within the Commonwealth, with Julius Nyerere as prime minister.
**1962** Tanganyika became a republic with Nyerere as president.
**1964** Tanganyika and Zanzibar became the United Republic of Tanzania with Nyerere as president.
**1967** East African Community (EAC) formed. Nyerere committed himself to building a socialist state through a series of village cooperatives (the Arusha Declaration).
**1977** Revolutionary Party of Tanzania (CCM) proclaimed the only legal party. EAC dissolved.
**1978** Ugandan forces repulsed after crossing into Tanzania.
**1979** Tanzanian troops sent to Uganda to help overthrow the president, Idi Amin.
**1985** Nyerere retired from presidency but stayed on as CCM leader; Ali Hassan Mwinyi became president.
**1990** Nyerere surrendered CCM leadership; replaced by President Mwinyi.
**1992** CCM agreed to abolish one-party rule. East African cooperation pact with Kenya and Uganda to be re-established.

## Thailand

**Thailand**
Kingdom of
(*Prathet Thai* or
*Muang-Thai*)
(formerly *Siam* to
1939 and 1945–49)

*area* 513,115 sq km/198,108 sq mi
*capital* Bangkok
*towns* Chiangmai, Nakhon Sawan
*physical* mountainous, semi-arid plateau in NE, fertile central region, tropical isthmus in S
*environment* tropical rainforest was reduced to 18% of the land area 1988 (from 93% in 1961); logging banned by the government 1988
*head of state* King Bhumibol Adulyadej from 1946
*head of government* Chuan Leekpai from 1992
*political system* military-controlled emergent democracy
*exports* rice, textiles, rubber, tin, rubies, sapphires, maize, tapioca
*currency* baht
*population* (1990 est) 54,890,000 (Thai 75%, Chinese 14%); growth rate 2% p.a.
*life expectancy* men 62, women 68 (1989)
*languages* Thai and Chinese (both official); regional dialects
*religions* Buddhist 95%, Muslim 4%
*literacy* 89% (1988)

***GNP*** $52 bn (1988); $771 per head (1988)
*chronology*
***1782*** Siam absolutist dynasty commenced.
***1896*** Anglo-French agreement recognized Siam as an independent buffer state.
***1932*** Constitutional monarchy established.
***1939*** Name of Thailand adopted.
***1941–44*** Japanese occupation.
***1947*** Military seized power in coup.
***1972*** Withdrawal of Thai troops from South Vietnam.
***1973*** Military government overthrown.
***1976*** Military reassumed control.
***1980*** General Prem Tinsulanonda assumed power.
***1983*** Civilian government formed; martial law maintained.
***1988*** Prime Minister Prem resigned; replaced by Chatichai Choonhavan.
***1989*** Thai pirates continued to murder, pillage, and kidnap Vietnamese 'boat people' at sea.
***1991*** Military seized power in coup. Interim civilian government formed under Anand Panyarachun. 50,000 demonstrated against new military-oriented constitution.
***1992*** March: general election produced five-party coalition; Narong Wongwan named premier but removed a month later. April: appointment of General Suchinda Kraprayoon as premier provoked widespread riots. May: Suchinda forced to stand down. June: Anand made interim prime minister. Sept: new coalition government led by Chuan Leekpai.

# Togo

**Togo**
Republic of
(*République Togolaise*)

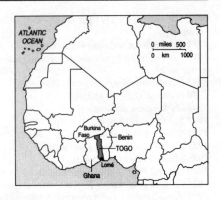

***area*** 56,800 sq km/21,930 sq mi
***capital*** Lomé
***towns*** Sokodé, Kpalimé
***physical*** two savanna plains, divided by range of hills NE–SW; coastal lagoons and marsh
***environment*** the homes of thousands of people in Keto were destroyed by coastal erosion as a result of the building of the Volta dam
***head of state*** Etienne Gnassingbé Eyadéma from 1967
***head of government*** Jospeh Kokou Koffigoh from 1991
***political system*** transitional
***exports*** phosphates, cocoa, coffee, coconuts
***currency*** franc CFA
***population*** (1990 est) 3,566,000; growth rate 3% p.a.
***life expectancy*** men 53, women 57 (1989)
***languages*** French (official), Ewe, Kabre
***religions*** animist 46%, Catholic 28%, Muslim 17%, Protestant 9%
***literacy*** men 53%, women 28% (1985 est)
***GNP*** $1.3 bn (1987); per head (1985)
***chronology***
*1885–1914* Togoland was a German protectorate until captured by Anglo-French forces.

*1922* Divided between Britain and France under League of Nations mandate.
*1946* Continued under United Nations trusteeship.
*1956* British Togoland integrated with Ghana.
*1960* French Togoland achieved independence from France as the Republic of Togo with Sylvanus Olympio as head of state.
*1963* Olympio killed in a military coup. Nicolas Grunitzky became president.
*1967* Grunitzky replaced by Lt-Gen Etienne Gnassingbé Eyadéma in bloodless coup.
*1973* Assembly of Togolese People (RPT) formed as sole legal political party.
*1975* Lomé convention signed in Lomé, establishing trade links between the European Community (EC) and developing countries.
*1979* Eyadéma returned in election. Further EC Lomé convention signed.
*1986* Attempted coup failed.
*1991* Eyadéma legalized opposition parties. National conference elected Joseph Kokou Koffigoh head of interim government; troops loyal to Eyadéma failed to reinstate him.
*1992* Overwhelming referendum support for multiparty politics.
*1993* Feb: all-party talks to avoid civil war began in France but were suspended after disagreements among participants.

# Tonga

**Tonga**
Kingdom of
(*Pule'anga Fakatu'i
'o Tonga*)
or *Friendly Islands*

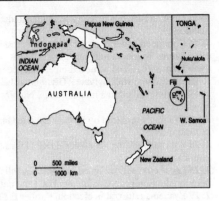

*area* 750 sq km/290 sq mi
*capital* Nuku'alofa (on Tongatapu island)
*towns* Pangai, Neiafu
*physical* three groups of islands in SW Pacific, mostly coral formations, but actively volcanic in W
*head of state* King Taufa'ahau Tupou IV from 1965
*head of government* Baron Vaea from 1991
*political system* constitutional monarchy
*currency* Tongan dollar or pa'anga
*population* (1988) 95,000; growth rate 2.4% p.a.
*life expectancy* men 69, women 74 (1989)
*languages* Tongan (official), English
*religions* Wesleyan 47%, Roman Catholic 14%, Free Church of Tonga 14%, Mormon 9%, Church of Tonga 9%
*literacy* 93% (1988)
*GNP* $65 million (1987); $430 per head
*chronology*
*1831* Tongan dynasty founded by Prince Taufa'ahau Tupou.
*1900* Became a British protectorate.
*1958* Treaty with Britain brought Tonga increased control over its internal affairs.

*1965* Queen Salote died; succeeded by her son, King Taufa'ahau Tupou IV.
*1967* Revised treaty with Britain brought greater independence.
*1970* Full independence achieved from Britain within the Commonwealth.
*1990* Three prodemocracy candidates elected. Calls for reform of absolutist power.

# Trinidad and Tobago
Republic of

*area* Trinidad 4,828 sq km/1,864 sq mi and Tobago 300 sq km/116 sq mi
*capital* Port-of-Spain
*towns* San Fernando, Arima, Scarborough (Tobago)
*physical* comprises two main islands and some smaller ones; coastal swamps and hills E–W
*head of state* Noor Hassanali from 1987
*head of government* Patrick Manning from 1991
*political system* democratic republic
*exports* oil, petroleum products, chemicals, sugar, cocoa
*currency* Trinidad and Tobago dollar
*population* (1990 est) 1,270,000 (African descent 40%, Indian 40%, European 16%, Chinese 2% and others), 1.2 million on Trinidad; growth rate 1.6% p.a.
*life expectancy* men 68, women 72 (1989)
*languages* English (official), Hindi, French, Spanish
*religions* Roman Catholic 32%, Protestant 29%, Hindu 25%, Muslim 6%
*literacy* 97% (1988)

***GNP*** $4.5 bn; $3,731 per head (1987)
*chronology*
***1888*** Trinidad and Tobago united as a British colony.
***1956*** People's National Movement (PNM) founded.
***1959*** Achieved internal self-government, with PNM leader Eric Williams as chief minister.
***1962*** Independence achieved from Britain, within the Commonwealth, with Williams as prime minister.
***1976*** Became a republic, with Ellis Clarke as president and Williams as prime minister.
***1981*** Williams died and was succeeded by George Chambers, with Arthur Robinson as opposition leader.
***1986*** National Alliance for Reconstruction (NAR), headed by Arthur Robinson, won general election.
***1987*** Noor Hassanali became president.
***1990*** Attempted antigovernment coup defeated.
***1991*** General election saw victory for PNM, with Patrick Manning as prime minister.

# Tunisia
Tunisian Republic
(*al-Jumhuriya at-Tunisiya*)

*area* 164,150 sq km/63,378 sq mi
*capital* Tunis
*towns* ports Sfax, Sousse, Bizerta
*physical* arable and forested land in N graduates towards desert in S
*head of state and government* Zine el-Abidine Ben Ali from 1987
*political system* emergent democratic republic
*exports* oil, phosphates, chemicals, textiles, food, olive oil
*currency* dinar
*population* (1990 est) 8,094,000; growth rate 2% p.a.
*life expectancy* men 68, women 71 (1989)
*languages* Arabic (official), French
*religion* Sunni Muslim 95%; Jewish, Christian
*literacy* men 68%, women 41% (1985 est)
*GNP* $9.6 bn (1987); $1,163 per head (1986)
*chronology*
*1883* Became a French protectorate.
*1955* Granted internal self-government.
*1956* Independence achieved from France as a monarchy, with Habib Bourguiba as prime minister.

*1957* Became a republic with Bourguiba as president.
*1975* Bourguiba made president for life.
*1985* Diplomatic relations with Libya severed.
*1987* Bourguiba removed Prime Minister Rashed Sfar and appointed Zine el-Abidine Ben Ali. Ben Ali declared Bourguiba incompetent and seized power.
*1988* Constitutional changes towards democracy announced. Diplomatic relations with Libya restored.
*1989* Government party, Constitutional Democratic Rally (RDC), won all assembly seats in general election.
*1991* Opposition to US actions during the Gulf War. Crackdown on religious fundamentalists.

## Turkey
Republic of
(*Türkiye Cumhuriyeti*)

*area* 779,500 sq km/300,965 sq mi
*capital* Ankara
*towns* Istanbul, Izmir
*physical* central plateau surrounded by mountains
*environment* only 0.3% of the country is protected by national parks and reserves compared with a global average of 7% per country
*head of state* to be appointed
*head of government* Suleyman Demirel from 1991
*political system* democratic republic
*exports* cotton, yarn, hazelnuts, citrus fruits, tobacco, dried fruit, chromium ores
*currency* Turkish lira
*population* (1990 est) 56,549,000 (Turkish 85%, Kurdish 12%); growth rate 2.1% p.a.
*life expectancy* men 63, women 66 (1989)
*languages* Turkish (official), Kurdish, Arabic
*religion* Sunni Muslim 98%
*literacy* men 86%, women 62% (1985)
*GNP* $62 bn (1987); $1,160 per head (1986)

## chronology

**1919–22** Turkish War of Independence provoked by Greek occupation of Izmir. Mustafa Kemal (Atatürk), leader of nationalist congress, defeated Italian, French, and Greek forces.
**1923** Treaty of Lausanne established Turkey as independent republic under Kemal. Westernization began.
**1950** First free elections; Adnan Menderes became prime minister.
**1960** Menderes executed after military coup by General Cemal Gürsel.
**1965** Suleyman Demirel became prime minister.
**1971** Army forced Demirel to resign.
**1973** Civilian rule returned under Bulent Ecevit.
**1974** Turkish troops sent to protect Turkish community in Cyprus.
**1975** Demirel returned to head of a right-wing coalition.
**1978** Ecevit returned, as head of coalition, in the face of economic difficulties and factional violence.
**1979** Demeril returned. Violence grew.
**1980** Army took over, and Bulent Ulusu became prime minister. Harsh repression of political activists attracted international criticism.
**1982** New constitution adopted.
**1983** Ban on political activity lifted. Turgut Özal became prime minister.
**1987** Özal maintained majority in general election.
**1988** Improved relations and talks with Greece.
**1989** Turgut Özal elected president; Yildirim Akbulut became prime minister. Application to join European Community (EC) rejected.
**1991** Mesut Yilmaz became prime minister. Turkey sided with UN coalition against Iraq in Gulf War. Conflict with Kurdish minority continued. Coalition government formed under Suleyman Demirel after inconclusive election result.
**1992** Earthquake claimed thousands of lives.
**1993** April: Turgut Özal died of heart attack.

# Turkmenistan
Republic of

*area* 488,100 sq km/188,406 sq mi
*capital* Ashkhabad
*towns* Chardzhov, Mary (Merv), Nebit-Dag, Krasnovodsk
*physical* some 90% of land is desert including the Kara Kum 'Black Sands' desert (area 310,800 sq km/120,000 sq mi)
*head of state* Saparmurad Niyazov from 1991
*head of government* Sakhat Muradov from 1992
*political system* socialist pluralist
*products* silk, karakul, sheep, astrakhan fur, carpets, chemicals, rich deposits of petroleum, natural gas, sulphur, and other industrial raw materials
*population* (1990) 3,600,000 (Turkmen 72%, Russian 10%, Uzbek 9%, Kazakh 3%, Ukrainian 1%)
*language* West Turkic, closely related to Turkish
*religion* Sunni Muslim
*chronology*
*1921* Part of Turkestan Soviet Socialist Autonomous Republic.

*1925* Became a constituent republic of USSR.
*1990* Aug: economic and political sovereignty declared.
*1991* Jan: Communist Party leader Sakhat Niyazov became state president. March: endorsed maintenance of the Union in USSR referendum. Aug: Niyazov initially supported attempted anti-Gorbachev coup. Oct: independence declared. Dec: joined new Commonwealth of Independent States (CIS).
*1992* Jan: admitted into Conference for Security and Cooperation in Europe (CSCE). Feb: joined the Muslim Economic Cooperation Organization (ECO). March: admitted into United Nations (UN); US diplomatic recognition achieved. May: new constitution adopted. Nov–Dec: 60-member parliament popularly elected with Sakhat Muradov as prime minister.

# Tuvalu
South West Pacific
State of
(formerly *Ellice Islands*)

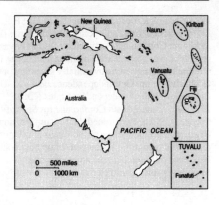

*area* 25 sq km/9.5 sq mi
*capital* Funafuti
*physical* nine low coral atolls forming a chain of 579 km/650 mi in the SW Pacific
*head of state* Elizabeth II from 1978, represented by governor general
*head of government* Bikenibeu Paeniu from 1989
*political system* liberal democracy
*exports* copra, handicrafts, stamps
*currency* Australian dollar
*population* (1990 est) 9,000 (Polynesian 96%); growth rate 3.4% p.a.
*life expectancy* 60 men, 63 women (1989)
*languages* Tuvaluan, English
*religion* Christian (Protestant)
*literacy* 96% (1985)
*GDP* (1983) $711 per head
*chronology*
*1892* Became a British protectorate forming part of the Gilbert and Ellice Islands group.
*1916* The islands acquired colonial status.
*1975* The Ellice Islands were separated from the Gilbert Islands.
*1978* Independence achieved from Britain, within the Commonwealth,

with Toaripi Lauti as prime minister.
*1981* Dr Tomasi Puapua replaced Lauti as premier; country reverted to former name Tuvalu.
*1986* Islanders rejected proposal for republican status.
*1989* Bikenibeu Paeniu elected new prime minister and pledged to reduce the country's dependence on foreign aid.

# Uganda
Republic of

*area* 236,600 sq km/91,351 sq mi
*capital* Kampala
*towns* Jinja, M'Bale, Entebbe, Masaka
*physical* plateau with mountains in W; forest and grassland; arid in NE
*head of state and government* Yoweri Museveni from 1986
*political system* emergent democratic republic
*exports* coffee, cotton, tea, copper
*currency* Uganda new shilling
*population* (1990 est) 17,593,000 (largely the Baganda, after whom the country is named; also Langi and Acholi, some surviving Pygmies); growth rate 3.3% p.a.
*life expectancy* men 49, women 51 (1989)
*languages* English (official), Kiswahili, Luganda, and other African languages
*religions* Roman Catholic 33%, Protestant 33%, Muslim 16%, animist
*literacy* men 70%, women 45% (1985 est)
*GNP* $3.6 bn (1987); $220 per head
*chronology*
*1962* Independence achieved from Britain, within the Commonwealth, with Milton Obote as prime minister.
*1963* Proclaimed a federal republic with King Mutesa II as president.

*1966* King Mutesa ousted in coup led by Obote, who ended the federal status and became executive president.
*1969* All opposition parties banned after an assassination attempt on Obote.
*1971* Obote overthrown in army coup led by Maj-Gen Idi Amin Dada; ruthlessly dictatorial regime established; nearly 49,000 Ugandan Asians expelled; over 300,000 opponents of regime killed.
*1978* Amin forced to leave country by opponents backed by Tanzanian troops. Provisional government set up with Yusuf Lule as president. Lule replaced by Godfrey Binaisa.
*1978–79* Fighting broke out against Tanzanian troops.
*1980* Binaisa overthrown by army. Elections held and Milton Obote returned to power.
*1985* After opposition by National Resistance Army (NRA), and indiscipline in army, Obote ousted by Brig Tito Okello; power-sharing agreement entered into with NRA leader Yoweri Museveni.
*1986* Agreement ended; Museveni became president, heading broad-based coalition government.
*1992* Announcement made that East African cooperation pact with Kenya and Tanzania would be revived.

# Ukraine

*area* 603,700 sq km/233,089 sq mi
*capital* Kiev
*towns* Kharkov, Donetsk, Odessa, Dnepropetrovsk, Lviv (Lvov), Krivoi Rog, Zaporozhye
*physical* Russian plain; Carpathian and Crimean Mountains; rivers: Dnieper (with the Dnieper dam 1932), Donetz, Bug
*head of state* Leonid Kravchuk from 1990
*head of government* Leonid Kuchma from 1992
*political system* emergent democracy
*products* grain, coal, oil, various minerals
*currency* grivna
*population* (1990) 51,800,000 (Ukrainian 73%, Russian 22%, Byelorussian, Russian-speaking Jews 1%—some 1.5 million have emigrated to the USA, 750,000 to Canada)
*language* Ukrainian (Slavonic), with a literature that goes back to the Middle Ages
*religions* traditionally Ukrainian Orthodox; also Ukrainian Catholic
*chronology*
*1918* Independent People's Republic proclaimed.

*1920* Conquered by Soviet Red Army.

*1921* Poland alloted charge of W Ukraine.

*1932–33* Famine caused the deaths of more than 7.5 million people.

*1939* W Ukraine occupied by Red Army.

*1941–44* Under Nazi control; Jews massacred at Babi Yar; more than five million Ukrainians and Ukrainian Jews deported and exterminated.

*1944* Soviet control re-established.

*1945* Became a founder member of the United Nations.

*1946* Ukrainian Uniate Church proscribed and forcibly merged with Russian Orthodox Church.

*1986* April: Chernobyl nuclear disaster.

*1989* Feb: Ukrainian People's Movement (Rukh) established. Ban on Ukrainian Uniate Church lifted.

*1990* July: voted to proclaim sovereignty; Leonid Kravchuk indirectly elected as president; sovereignty declared.

*1991* Aug: demonstrations against the abortive anti-Gorbachev coup; independence declared, pending referendum; Communist Party (CP) activities were suspended. Oct: voted to create independent army. Dec: Kravchuk popularly elected president; independence overwhelmingly endorsed in referendum; joined new Commonwealth of Independent States (CIS); independence acknowledged by USA and European Community (EC).

*1992* Jan: admitted into Conference on Security and Cooperation in Europe (CSCE); pipeline deal with Iran to end dependence on Russian oil; prices freed. Feb: prices 'temporarily' re-regulated. March: agreed tactical arms-shipments to Russia suspended. May: Crimean sovereignty declared, but subsequently rescinded. Aug: joint control of Black Sea fleet agreed with Russia. Oct: Leonid Kuchma became prime minister. Production declined by 20% during 1992.

*1993* Inflation at 35% a month in early part of year; budget deficit at 44% of GDP.

## United Arab Emirates
(UAE)
(*Ittihad al-Imarat al-Arabiyah*) federation of the emirates of Abu Dhabi, Ajman, Dubai, Fujairah, Ras al Khaimah, Sharjah, Umm al Qaiwain

*total area* 83,657 sq km/32,292 sq mi
*capital* Abu Dhabi
*towns* Dubai
*physical* desert and flat coastal plain; mountains in E
*head of state and of government* Sheik Sultan Zayed bin al-Nahayan of Abu Dhabi from 1971
*political system* absolutism
*exports* oil, natural gas, fish, dates
*currency* UAE dirham
*population* (1990 est) 2,250,000 (10% nomadic); growth rate 6.1% p.a.
*life expectancy* men 68, women 72 (1989)
*languages* Arabic (official), Farsi, Hindi, Urdu, English
*religions* Muslim 96%, Christian, Hindu
*literacy* 68% (1989)
*GNP* $22 bn (1987); $11,900 per head
*chronology*
*1952* Trucial Council established.
*1971* Federation of Arab Emirates formed; later dissolved. Six Trucial States formed United Arab Emirates, with ruler of Abu Dhabi, Sheik

Zayed, as president.
*1972* The seventh state, Ras al Khaimah, joined the federation.
*1976* Sheik Zayed threatened to relinquish presidency unless progress towards centralization became more rapid.
*1985* Diplomatic and economic links with USSR and China established.
*1987* Diplomatic relations with Egypt restored.
*1990–91* Iraqi invasion of Kuwait opposed; UAE fights with the UN coalition.
*1991* Bank of Commerce and Credit International (BCCI), controlled by Abu Dhabi's ruler, collapsed.

# United Kingdom

**United Kingdom** of Great Britain and Northern Ireland (UK)

*area* 244,100 sq km/94,247 sq mi
*capital* London
*towns* Birmingham, Glasgow, Leeds, Liverpool, Manchester, Belfast, Cardiff
*physical* became separated from European continent about 6000 BC; rolling landscape, increasingly mountainous towards the N, with Grampian Mountains in Scotland, Pennines in N England, Cambrian Mountains in Wales; rivers include Thames, Severn, and Spey
*territories* Anguilla, Bermuda, British Antarctic Territory, British Indian Ocean Territory, British Virgin Islands, Cayman Islands, Falkland Islands, Gibraltar, Hong Kong (until 1997), Montserrat, Pitcairn Islands, St Helena and Dependencies (Ascension, Tristan da Cunha), Turks and Caicos Islands
*environment* an estimated 67% (the highest percentage in Europe) of forests have been damaged by acid rain
*head of state* Elizabeth II from 1952
*head of government* John Major from 1990
*political system* liberal democracy

*exports* cereals, rape, sugar beet, potatoes, meat and meat products, poultry, dairy products, electronic and telecommunications equipment, engineering equipment and scientific instruments, oil and gas, petrochemicals, pharmaceuticals, fertilizers, film and television programmes, aircraft
*currency* pound sterling
*population* (1990 est) 57,121,000 (English 81.5%, Scottish 9.6%, Welsh 1.9%, Irish 2.4%, Ulster 1.8%); growth rate 0.1% p.a.
*religion* Christian (Protestant 55%, Roman Catholic 10%); Muslim, Jewish, Hindu, Sikh
*life expectancy* men 72, women 78 (1989)
*languages* English, Welsh, Gaelic
*literacy* 99% (1989)
*GNP* $758 bn; $13,329 per head (1988) 7z
*recent chronology*
*1945* Labour government under Clement Attlee; welfare state established.
*1951* Conservatives under Winston Churchill defeated Labour.
*1956* Suez Crisis.
*1964* Labour victory under Harold Wilson.
*1970* Conservatives under Edward Heath defeated Labour.
*1972* Parliament prorogued in Northern Ireland; direct rule from Westminster began.
*1973* UK joined European Economic Community.
*1974* Three-day week, coal strike; Wilson replaced Heath.
*1976* James Callaghan replaced Wilson as prime minister.
*1977* Liberal–Labour pact.
*1979* Victory for Conservatives under Margaret Thatcher.
*1981* Formation of Social Democratic Party (SDP). Riots occurred in inner cities.
*1982* Unemployment over 3 million. Falklands War.
*1983* Thatcher re-elected.
*1984–85* Coal strike, the longest in British history.
*1986* Abolition of metropolitan counties.
*1987* Thatcher re-elected for third term.
*1988* Liberals and most of SDP merged into the Social and Liberal

Democrats, leaving a splinter SDP. Inflation and interest rates rose.
*1989* The Green Party polled 2 million votes in the European elections.
*1990* Riots as poll tax introduced in England. Troops sent to the Persian Gulf following Iraq's invasion of Kuwait. British hostages held in Iraq, later released. Britain joined European exchange rate mechanism (ERM). Thatcher replaced by John Major as Conservative leader and prime minister.
*1991* British troops took part in US-led war against Iraq under United Nations umbrella. Support was given to the USSR during the dissolution of communism and the restoration of independence to the republics. John Major visited Beijing to sign agreement with China on new Hong Kong airport. At home, Britain suffered severe economic recession and rising unemployment.
*1992* Economic recession continued. April: Conservative Party, led by John Major, won fourth consecutive general election, but with reduced majority. Neil Kinnock resigned as Labour leader. July: John Smith became new Labour leader. Sept: sterling devalued and UK withdrawn from ERM. Oct: drastic pit-closure programme encountered massive public opposition; subsequently reviewed. John Major's popularity at unprecedentedly low rating. Nov: government motion in favour of ratification of Maastricht Treaty narrowly passed. Revelations of past arms sales to Iraq implicated senior government figures, including the prime minister.

# United States of America

***area*** 9,368,900 sq km/3,618,770 sq mi
***capital*** Washington DC
***towns*** New York, Los Angeles, Chicago, Philadelphia, Detroit, San Francisco
***physical*** topography and vegetation from tropical (Hawaii) to arctic (Alaska); mountain ranges parallel with E and W coasts; the Rocky Mountains separate rivers emptying into the Pacific from those flowing into the Gulf of Mexico; Great Lakes in N; rivers include Hudson, Mississippi, Missouri, Colorado, Columbia, Snake, Rio Grande, Ohio
***environment*** the USA produces the world's largest quantity of municipal waste per person (850 kg/1,900 lb)
***territories*** the commonwealths of Puerto Rico and Northern Marianas; the federated states of Micronesia; Guam, the US Virgin Islands, American Samoa, Wake Island, Midway Islands, Marshall Islands, Belau, and Johnston and Sand Islands

# United States of America

***head of state and government*** Bill Clinton from 1993
***political system*** liberal democracy
***currency*** US dollar
***population*** (1990 est) 250,372,000 (white 80%, black 12%, Asian/Pacific islander 3%, American Indian, Inuit, and Aleut 1%, Hispanic (included in above percentages) 9%); growth rate 0.9% p.a.
***life expectancy*** men 72, women 79 (1989)
***languages*** English, Spanish
***religions*** Christian 86.5% (Roman Catholic 26%, Baptist 19%, Methodist 8%, Lutheran 5%), Jewish 1.8%, Muslim 0.5%, Buddhist and Hindu less than 0.5%
***literacy*** 99% (1989)
***GNP*** $3,855 bn (1983); $13,451 per head
***recent chronology***
***1945*** USA ended war in the Pacific by dropping atom bombs on Hiroshima and Nagasaki, Japan.
***1950–53*** US involvement in Korean War. McCarthy anticommunist investigations (HUAC) became a 'witch hunt'.
***1954*** Civil Rights legislation began with the ending of the segregation of black and white students in public schools.
***1957*** Civil Rights bill on voting.
***1958*** First US satellite in orbit.
***1961*** Abortive CIA-backed invasion of Cuba at the Bay of Pigs.
***1963*** President Kennedy assassinated; L B Johduring nson assumed the presidency.
***1964–68*** 'Great Society' civil-rights and welfare measures in the Omnibus Civil Rights bill.
***1964–75*** US involvement in Vietnam War.
***1965*** US intervention in Dominican Republic.
***1969*** US astronaut Neil Armstrong was the first human on the Moon.
***1973*** OPEC oil embargo almost crippled US industry and consumers. Inflation began.
***1973–74*** Watergate scandal began during Richard Nixon's re-election campaign and ended just before impeachment; Nixon resigned as president; replaced by Gerald Ford, who 'pardoned' Nixon.
***1975*** Final US withdrawal from Vietnam.

***1979*** US–Chinese diplomatic relations normalized.
***1981*** Ronald Reagan inaugurated as president. Space shuttle mission was successful.
***1983*** US invasion of Grenada.
***1986*** 'Irangate' scandal over secret US government arms sales to Iran, with proceeds to antigovernment Contra guerrillas in Nicaragua.
***1987*** Reagan and Gorbachev (for USSR) signed intermediate-range nuclear forces treaty. Wall Street stock-market crash caused by programme trading.
***1988*** USA became world's largest debtor nation, owing $532 billion. George Bush elected president.
***1989*** Bush met Gorbachev at Malta, end to Cold War declared; high-level delegation sent to China amid severe criticism; large troop reductions and budget cuts announced for US military; USA invaded Panama; Noriega taken into custody.
***1990*** Bush and Gorbachev met again. Nelson Mandela freed in South Africa, toured USA. US troops sent to Middle East following Iraq's invasion of Kuwait.
***1991*** Jan–Feb: US-led assault drove Iraq from Kuwait in Gulf War. US support was given to the USSR during the dissolution of communism and the recognition of independence of the Baltic republics. July: Strategic Arms Reduction Treaty (START) signed at US–Soviet summit in Moscow. Nov: Bush co-hosted Middle East peace conference in Spain.
***1992*** Bush's popularity slumped as economic recession continued. Widespread riots in Los Angeles. Nov: Bill Clinton won presidential elections for the Democrats; independent candidate Ross Perot won nearly 20% of votes.
***1993*** Jan: Clinton delayed executive order to suspend ban on homosexuality in the armed forces. Feb: medium-term economic plan proposed by Clinton to Congress to cut federal budget deficit.

## Uruguay
Oriental Republic of
(*República Oriental del Uruguay*)

*area* 176,200 sq km/68,031 sq mi
*capital* Montevideo
*towns* Salto, Paysandú
*physical* grassy plains (pampas) and low hills
*head of state and government* Luis Lacalle Herrera from 1989
*political system* democratic republic
*exports* meat and meat products, leather, wool, textiles
*currency* nuevo peso
*population* (1990 est) 3,002,000 (Spanish, Italian; mestizo, mulatto, black); growth rate 0.7% p.a.
*life expectancy* men 68, women 75 (1989)
*language* Spanish
*religion* Roman Catholic 66%
*literacy* 96% (1984)
*GNP* $7.5 bn; $2,470 per head (1988)

*chronology*
*1825* Independence declared from Brazil.
*1836* Civil war.
*1930* First constitution adopted.
*1966* Blanco party in power, with Jorge Pacheco Areco as president.
*1972* Colorado Party returned, with Juan Maria Bordaberry Arocena as president.
*1976* Bordaberry deposed by army; Dr Méndez Manfredini became president.
*1984* Violent antigovernment protests after ten years of repressive rule.
*1985* Agreement reached between the army and political leaders for return to constitutional government. Colorado Party won general election; Dr Julio Maria Sanguinetti became president.
*1986* Government of national accord established under President Sanguinetti's leadership.
*1989* Luis Lacalle Herrera elected president.

# Uzbekistan
Republic of

*area* 447,400 sq km/172,741 sq mi
*capital* Tashkent
*towns* Samarkand, Bukhara, Namangan
*physical* oases in the deserts; rivers: Amu Darya, Syr Darya; Fergana Valley; rich in mineral deposits
*head of state* Islam Karimov from 1990
*head of government* Abdul Hashim Mutalov from 1991
*political system* socialist pluralist
*products* rice, dried fruit, vines (all grown by irrigation); cotton, silk
*population* (1990) 20,300,000 (Uzbek 71%, Russian 8%, Tajik 5%, Kazakh 4%)
*language* Uzbek, a Turkic language
*religion* Sunni Muslim
*chronology*
*1921* Part of Turkestan Soviet Socialist Autonomous Republic.
*1925* Became constituent republic of the USSR.
*1944* Some 160,000 Meskhetian Turks forcibly transported from their

native Georgia to Uzbekistan by Stalin.
*1989* June: Tashlak, Yaipan, and Ferghana were the scenes of riots in which Meskhetian Turks were attacked; 70 killed and 850 wounded.
*1990* June: economic and political sovereignty declared; Islam Karimov became president.
*1991* March: Uzbek supported 'renewed federation' in USSR referendum. Aug: anti-Gorbachev coup in Moscow initially accepted by President Karimov; later, Karimov resigned from Soviet Communist Party (CPSU) Politburo; Uzbek Communist Party (UCP) broke with CPSU; pro-democracy rallies dispersed by militia; independence declared. Dec: joined new Commonwealth of Independent States (CIS).
*1992* Jan: admitted into Conference on Security and Cooperation in Europe (CSCE); violent food riots in Tashkent. March: joined the United Nations (UN); US diplomatic recognition achieved.

# Vanuatu
Republic of
(*Ripablik Blong Vanuatu*)

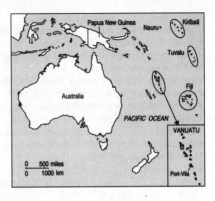

*area* 14,800 sq km/5,714 sq mi
*capital* Vila, or Port-Vila, (on Efate)
*towns* Luganville (on Espíritu Santo)
*physical* comprises around 70 islands, including Espíritu Santo, Malekula, and Efate; densely forested, mountainous
*head of state* Fred Timakata from 1989
*head of government* Maxime Carlot from 1991
*political system* democratic republic
*exports* copra, fish, coffee, cocoa
*currency* vatu
*population* (1989) 152,000 (90% Melanesian); growth rate 3.3% p.a.
*life expectancy* men 67, women 71 (1989)
*languages* Bislama 82%, English, French (all official)
*literacy* 53%
*religions* Presbyterian 40%, Roman Catholic 16%, Anglican 14%, animist 15%
*GDP* $125 million (1987); $927 per head
*chronology*
*1906* Islands jointly administered by France and Britain.
*1975* Representative assembly established.
*1978* Government of national unity formed, with Father Gerard

Leymang as chief minister.
*1980* Revolt on the island of Espíritu Santo delayed independence but it was achieved within the Commonwealth, with George Kalkoa (adopted name Sokomanu) as president and Father Walter Lini as prime minister.
*1988* Dismissal of Lini by Sokomanu led to Sokomanu's arrest for treason. Lini reinstated.
*1989* Sokomanu sentenced to six years' imprisonment; succeeded as president by Fred Timakata.
*1991* Lini voted out by party members; replaced by Donald Kalpokas. General election produced Union of Moderate Parties–Vanuatu National Party coalition under Maxime Carlot.

# Vatican City State
(*Stato della Città del Vaticano*)

*area* 0.4 sq km/109 acres
*physical* forms an enclave in the heart of Rome, Italy
*head of state and government* John Paul II from 1978
*political system* absolute Catholicism
*currency* Vatican City lira; Italian lira
*population* (1985) 1,000
*languages* Latin (official), Italian
*religion* Roman Catholic
*chronology*
*1929* Lateran Treaty recognized sovereignty of the pope.
*1947* New Italian constitution confirmed the sovereignty of the Vatican City State.
*1978* John Paul II became the first non-Italian pope for more than 400 years.
*1985* New concordat signed under which Roman Catholicism ceased to be Italy's state religion.

# Venezuela
Republic of
(*República de Venezuela*)

***area*** 912,100 sq km/352,162 sq mi
***capital*** Caracas
***towns*** Barquisimeto, Valencia, Maracaibo
***physical*** Andes Mountains and Lake Maracaibo in NW; central plains (llanos); delta of river Orinoco in E; Guiana Highlands in SE
***head of state and of government*** Carlos Andrés Pérez from 1988
***government*** federal democratic republic
***exports*** coffee, timber, oil, aluminium, iron ore, petrochemicals
***currency*** bolívar
***population*** (1990 est) 19,753,000 (mestizos 70%, white (Spanish, Portuguese, Italian) 20%, black 9%, Amerindian 2%); growth rate 2.8% p.a.
***life expectancy*** men 67, women 73 (1989)
***religions*** Roman Catholic 96%, Protestant 2%
***languages*** Spanish (official), Amerindian languages 2%
***literacy*** 88% (1989)
***GNP*** $47.3 bn (1988); $2,629 per head (1985)
***recent chronology***
***1961*** New constitution adopted, with Rómulo Betancourt as president.

*1964* Dr Raúl Leoni became president.
*1969* Dr Rafael Caldera became president.
*1974* Carlos Andrés Pérez became president.
*1979* Dr Luis Herrera became president.
*1984* Dr Jaime Lusinchi became president; social pact established between government, trade unions, and business; national debt rescheduled.
*1987* Widespread social unrest triggered by inflation; student demonstrators shot by police.
*1988* Carlos Andrés Pérez elected president. Payments suspended on foreign debts (increase due to drop in oil prices).
*1989* Economic austerity programme enforced by $4.3 billion loan from International Monetary Fund (IMF). Price increases triggered riots; 300 people killed. Feb: martial law declared. May: General strike. Elections boycotted by opposition groups.
*1991* Protests against austerity programme continued.
*1992* Attempted anti-government coups failed. Pérez promised constitutional changes.

# Vietnam
Socialist Republic of
(*Công Hòa Xã Hôi Chu Nghĩa Viêt Nam*)

***area*** 329,600 sq km/127,259 sq mi
***capital*** Hanoi
***towns*** ports Ho Chi Minh City (formerly Saigon), Da Nang, Haiphong
***physical*** Red River and Mekong deltas, centre of cultivation and population; tropical rainforest; mountainous in N and NW
***environment*** during the Vietnam War an estimated 2.2 million hectares/5.4 million acres of forest were destroyed. The country's National Conservation Strategy is trying to replant 500 million trees each year
***head of state*** Le Duc Anh from 1992
***head of government*** Vo Van Kiet from 1991
***political system*** communism
***exports*** rice, rubber, coal, iron, apatite
***currency*** dong
***population*** (1990 est) 68,488,000 (750,000 refugees, mainly ethnic Chinese, left 1975–79; some settled in China, others fled by sea—the 'boat people'—to Hong Kong and elsewhere); growth rate 2.4% p.a.

*life expectancy* men 62, women 66 (1989)
*languages* Vietnamese (official), French, English, Khmer, Chinese, local languages
*religions* Buddhist, Taoist, Confucian, Christian
*literacy* 78% (1989)
*GNP* $12.6 bn; $180 per head (1987)
*chronology*
*1945* Japanese removed from Vietnam at end of World War II.
*1946* Commencement of Vietminh war against French.
*1954* France defeated at Dien Bien Phu. Vietnam divided along 17th parallel.
*1964* US troops entered Vietnam War.
*1973* Paris cease-fire agreement.
*1975* Saigon captured by North Vietnam.
*1976* Socialist Republic of Vietnam proclaimed.
*1978* Admission into Comecon. Vietnamese invasion of Cambodia.
*1979* Sino-Vietnamese border war.
*1986* Retirement of 'old guard' leaders.
*1987–88* Over 10,000 political prisoners released.
*1988–89* Troop withdrawals from Cambodia continued.
*1989* 'Boat people' leaving Vietnam murdered and robbed at sea by Thai pirates. Troop withdrawal from Cambodia completed. Hong Kong forcibly repatriated some Vietnamese refugees.
*1991* Vo Van Kiet replaced Do Muoi as prime minister.
Cambodia peace agreement signed. Relations with China normalized.
*1992* Sept: Le Duc Anh elected president. Dec: relations with South Korea normalized; USA eased 30-year-old trade embargo.

# Yemen
Republic of
(*al Jamhuriya al Yamaniya*)

*area* 531,900 sq km/205,367 sq mi
*capital* San'a
*towns* Ta'iz, Aden
*physical* hot moist coastal plain, rising to plateau and desert
*head of state and of government* Ali Abdullah Saleh from 1990
*political system* authoritarian republic
*exports* cotton, coffee, grapes, vegetables
*currency* rial
*population* (1990 est) 11,000,000; growth rate 2.7% p.a.
*life expectancy* men 47, women 50
*language* Arabic
*religions* Sunni Muslim 63%, Shi'ite Muslim 37%
*literacy* men 20%, women 3% (1985 est)
*GNP* $4.9 bn (1983); $520 per head
*chronology*
*1918* Yemen became independent.
*1962* North Yemen declared the Yemen Arab Republic (YAR), with Abdullah al-Sallal as president. Civil war broke out between royalists and republicans.

*1967* Civil war ended with the republicans victorious. Sallal deposed and replaced by Republican Council. The People's Republic of South Yemen was formed.

*1970* People's Republic of South Yemen renamed People's Democratic Republic of Yemen.

*1971–72* War between South Yemen and the YAR; union agreement signed but not kept.

*1974* Ibrahim al-Hamadi seized power in North Yemen and Military Command Council set up.

*1977* Hamadi assassinated and replaced by Ahmed ibn Hussein al-Ghashmi.

*1978* Constituent People's Assembly appointed in North Yemen and Military Command Council dissolved. Ghashmi killed by envoy from South Yemen; succeeded by Ali Abdullah Saleh. War broke out again between the two Yemens. South Yemen president deposed and Yemen Socialist Party (YSP) formed with Abdul Fattah Ismail as secretary general, later succeeded by Ali Nasser Muhammad.

*1979* Cease-fire agreed with commitment to future union.

*1983* Saleh elected president of North Yemen for a further five-year term.

*1984* Joint committee on foreign policy for the two Yemens met in Aden.

*1985* Ali Nasser re-elected secretary general of the YSP in South Yemen; removed his opponents. Three bureau members were killed.

*1986* Civil war in South Yemen; Ali Nasser dismissed. New administration under Haydar Abu Bakr al-Attas.

*1988* President Saleh re-elected in North Yemen.

*1989* Draft constitution for single Yemen state published.

*1990* Border between two Yemens opened; countries formally united 22 May as Republic of Yemen.

*1991* New constitution approved.

*1992* Anti-government riots.

*1993* Multiparty parliamentary elections set for April.

# Yugoslavia

*area* 88,400 sq km/34,100 sq mi
*capital* Belgrade
*towns* Kraljevo, Leskovac, Pristina, Novi Sad, Titograd
*physical* mountainous, with river Danube plains in N and E; limestone (Karst) features in NW
*head of state* Dobrica Cosic from 1992
*head of government* Radoje Kontic from 1993
*political system* socialist pluralist republic
*exports* machinery, electrical goods, chemicals, clothing, tobacco
*currency* dinar
*population* (1990) 12,420,000 (Serb 53%, Albanian 15%, Macedonian 11%, Montenegrin 5%, Muslim 3%, Croat 2%)
*life expectancy* men 69, women 75 (1989)
*languages* Serbian variant of Serbo-Croatian, Macedonian, Slovenian
*religion* Eastern Orthodox 41% (Serbs), Roman Catholic 12% (Croats), Muslim 3%
*literacy* 90% (1989)
*GNP* $154.1 bn; $6,540 per head (1988)

*chronology*
*1918* Creation of Kingdom of the Serbs, Croats, and Slovenes.
*1929* Name of Yugoslavia adopted.
*1941* Invaded by Germany.
*1945* Yugoslav Federal Republic formed under leadership of Tito; communist constitution introduced.
*1948* Split with USSR.
*1961* Nonaligned movement formed under Yugoslavia's leadership.
*1980* Tito died; collective leadership assumed power.
*1988* Economic difficulties: 1,800 strikes, 250% inflation, 20% unemployment. Ethnic unrest in Montenegro and Vojvodina; party reshuffled and government resigned.
*1989* Reformist Croatian Ante Marković became prime minister. 29 died in ethnic riots in Kosovo province, protesting against Serbian attempt to end autonomous status of Kosovo and Vojvodina; state of emergency imposed. May: inflation rose to 490%; tensions with ethnic Albanians rose.
*1990* Multiparty systems established in Serbia and Croatia.
*1991* June: Slovenia and Croatia declared independence, resulting in clashes between federal and republican armies; Slovenia accepted EC-sponsored peace pact. Fighting continued in Croatia; repeated calls for cease-fires failed. Dec: President Stipe Mesic and Prime Minister Ante Marković resigned.
*1992* Jan: EC-brokered cease-fire established in Croatia; EC and USA recognized Slovenia's and Croatia's independence. Bosnia-Herzegovina and Macedonia declared independence. April: Bosnia-Herzegovina recognized as independent by EC and USA amid increasing ethnic hostility; bloody civil war ensued. New Federal Republic of Yugoslavia (FRY) proclaimed by Serbia and Montenegro but not recognized externally. May: Western ambassadors left Belgrade. International sanctions imposed against Serbia and Montenegro. Hostilities continued. Jun: Dobrica Cosic became president. Jul: Milan Panic became prime minister. Sept: UN membership suspended. Dec: Slobodan Milosevic re-elected Serbian president; Panic removed from office in vote of no confidence.
*1993* Radoje Kontic became prime minister.

## Zaire
Republic of
(*République du Zaïre*)
(formerly *Congo*)

*area* 2,344,900 sq km/905,366 sq mi
*capital* Kinshasa
*towns* Lubumbashi, Kananga, Kisangani, Matadi, Boma
*physical* Zaïre River basin has tropical rainforest and savanna; mountains in E and W
*head of state* Mobutu Sese Seko Kuku Ngbendu wa Zabanga
*head of government* Faustin Birindwa from 1993
*political system* socialist pluralist republic
*exports* coffee, copper, cobalt (80% of world output), industrial diamonds, palm oil
*currency* zaïre
*population* (1990 est) 35,330,000; growth rate 2.9% p.a.
*life expectancy* men 51, women 54 (1989)
*languages* French (official), Swahili, Lingala, other African languages; over 300 dialects
*religions* Christian 70%, Muslim 10%
*literacy* men 79%, women 45% (1985 est)

**GNP** $5 bn (1987); $127 per head
*chronology*
*1908* Congo Free State annexed to Belgium.
*1960* Independence achieved from Belgium as Republic of the Congo. Civil war broke out between central government and Katanga province.
*1963* Katanga war ended.
*1967* New constitution adopted.
*1970* Col Mobutu elected president.
*1971* Country became the Republic of Zaire.
*1972* The Popular Movement of the Revolution (MPR) became the only legal political party. Katanga province renamed Shaba.
*1974* Foreign-owned businesses and plantations seized by Mobutu and given in political patronage.
*1977* Original owners of confiscated properties invited back. Mobutu re-elected; Zairians invaded Shaba province from Angola, repulsed by Belgian paratroopers.
*1978* Second unsuccessful invasion from Angola.
*1988* Potential rift with Belgium avoided.
*1990* Mobutu announced end of ban on multiparty politics, following internal dissent.
*1991* Multiparty elections promised. Sept: after antigovernment riots, Mobutu agreed to share power with opposition; Etienne Tshisekedi appointed premier. Oct: Tshisekedi dismissed.
*1992* Aug: Tshisekedi reinstated against Mobuto's wishes; interim opposition parliament formed. Oct: renewed rioting.
*1993* Jan: French ambassador shot dead by loyalist troops during army mutiny; France and Belgium prepared to evacuate civilians. March: Tshisekedi dismissed by Mobutu, replaced by Faustin Birindwa, but Tshisekedi still considered himself in office.

# Zambia
Republic of

*area* 752,600 sq km/290,579 sq mi
*capital* Lusaka
*towns* Kitwe, Ndola, Kabwe, Chipata, Livingstone
*physical* forested plateau cut through by rivers
*head of state and government* Frederick Chiluba from 1991
*political system* socialist pluralist republic
*exports* copper, cobalt, zinc, emeralds, tobacco
*currency* kwacha
*population* (1990 est) 8,119,000; growth rate 3.3% p.a.
*life expectancy* men 54, women 57 (1989)
*language* English (official); Bantu dialects
*religions* Christian 66%, animist, Hindu, Muslim
*literacy* 54% (1988)
*GNP* $2.1 bn (1987); $304 per head (1986)
*chronology*
*1899–1924* As Northern Rhodesia, under administration of the British South Africa Company.

# Zambia

*1924* Became a British protectorate.
*1964* Independence achieved from Britain, within the Commonwealth, as the Republic of Zambia with Kenneth Kaunda as president.
*1972* United National Independence Party (UNIP) declared the only legal party.
*1976* Support for the Patriotic Front in Rhodesia declared.
*1980* Unsuccessful coup against President Kaunda.
*1985* Kaunda elected chair of the African Front Line States.
*1987* Kaunda elected chair of the Organization of African Unity (OAU).
*1988* Kaunda re-elected unopposed for sixth term.
*1990* Multiparty system announced for 1991.
*1991* Movement for Multiparty Democracy won landslide election victory; Frederick Chiluba became president.
*1992* Food and water shortages caused by severe drought.

# Zimbabwe
Republic of

*area* 390,300 sq km/150,695 sq mi
*capital* Harare
*towns* Bulawayo, Gweru, Kwekwe, Mutare, Hwange
*physical* high plateau with central high veld and mountains in E; rivers Zambezi, Limpopo
*head of state and government* Robert Mugabe from 1987
*political system* effectively one-party socialist republic
*exports* tobacco, asbestos, cotton, coffee, gold, silver, nickel, copper
*currency* Zimbabwe dollar
*population* (1990 est) 10,205,000 (Shona 80%, Ndbele 19%; about 100,000 whites); growth rate 3.5% p.a.
*life expectancy* men 59, women 63 (1989)
*languages* English (official), Shona, Sindebele
*religions* Christian, Muslim, Hindu, animist
*literacy* men 81%, women 67% (1985 est)
*GNP* $5.5 bn (1988); $275 per head (1986)

*chronology*
*1889–1923* As Southern Rhodesia, under administration of British South Africa Company.
*1923* Became a self-governing British colony.
*1961* Zimbabwe African People's Union (ZAPU) formed, with Joshua Nkomo as leader; declared illegal 1962.
*1963* Zimbabwe African National Union (ZANU) formed, with Robert Mugabe as secretary general.
*1964* Ian Smith became prime minister. ZANU banned. Nkomo and Mugabe imprisoned.
*1965* Smith declared unilateral independence.
*1974* Nkomo and Mugabe released.
*1975* Geneva conference set date for constitutional independence.
*1979* Smith produced new constitution and established a government with Bishop Abel Muzorewa as prime minister. New government denounced by Nkomo and Mugabe. Conference in London agreed independence arrangements (Lancaster House Agreement).
*1980* Independence achieved from Britain, with Robert Mugabe as prime minister and Rev Canaan Banana as president.
*1982* Nkomo dismissed from the cabinet, leaving the country temporarily.
*1984* ZANU–Patriotic Front (PF) party congress agreed to create a one-party state in future.
*1985* Relations between Mugabe and Nkomo improved. Troops sent to Matabeleland to suppress rumoured insurrection.
*1987* Separate assembly seats for whites abolished. President Banana retired; Mugabe combined posts of head of state and prime minister and took on the title executive president.
*1988* Nkomo returned to the cabinet and was appointed vice president.
*1989* Opposition party, the Zimbabwe Unity Movement, formed by Edgar Tekere; draft constitution drawn up, renouncing Marxism–Leninism; ZANU–PF and ZAPU formally merged.
*1990* ZANU–PF re-elected. State of emergency ended. Opposition to creation of one-party state.
*1992* United Front formed to oppose ZANU–PF. March: Mugabe declared dire drought and famine situation a national disaster